5G AND NEXT-GEN CONSUMER BANKING SERVICES

T0299232

5G AND NEXT-GEN CONSUMER BANKING SERVICES

SC. Krunoslav Ris

CRC Press
Taylor & Francis Group
Boca Raton London New York

CRC Press is an imprint of the
Taylor & Francis Group, an **informa** business

First edition published 2022
by CRC Press
6000 Broken Sound Parkway NW, Suite 300, Boca Raton, FL 33487-2742

and by CRC Press
4 Park Square, Milton Park, Abingdon, Oxon, OX14 4RN

CRC Press is an imprint of Taylor & Francis Group, LLC

ISBN: 978-1-032-05572-5 (hbk)
ISBN: 978-1-032-05574-9 (pbk)
ISBN: 978-1-003-19817-8 (ebk)

DOI: 10.1201/9781003198178

Typeset in Caslon
by SPi Technologies India Pvt Ltd (Straive)

Table of Contents

Acknowledgements

Writing a book has always been my desire, somehow in the past, I knew that one day something would come out of me. But, as for every great thing a man does, it is necessary to have patience and understanding of the environment, and above all, those who are with you and who encourage you when you give up or find yourself in a creative vacuum.

My inspiration and the people who pushed me and encouraged me are my beloved wife Nelli, whom I thank endlessly for understanding during the months of work, my mother, who immensely supports me in all my craziness. I also want to thank my best friends Goran and Darko, who were with me every time I fell to pick me up so I can continue full sails forward.

I thank all of them for their presence and I wish to share with them many more good things in the future.

Author

Dr. SC. **Krunoslav Ris** was born and raised in the 1980s, when the internet was a Sci-Fi and Personal Computer rarity. It was the age of Pac-Man, Space Invaders, Commodore 64, and the programming language BASIC.

Kruno (for friends and family) is in IT business for more than 20 years and he has spent the last eight years in FinTech industry where he was directly involved in projects as Project Manager/Solution Architect/Team Leader/CTO/CEO/Head of Segment. He gathered experience working on more than 400 projects in his career and become a Croatian Futurist and speaker, author, and media personality who covered the future of the business and ongoing Digital Transformation.

He is the founder of several start-up companies. In 2020, he earned his Ph.D. with a dissertation on A.I. in FinTech and next-gen consumer banking. He also lectures on FinTech, Digital Disruption, Digital Transformation, and Blockchain at several Southern Eastern Europe Universities.

The book, *5G and Next-Gen Consumer Banking* was written as a result of three years of work in Direct Channels division in one of the major Croatian Banks in which he saw what will be the future trends when consumers become mature.

1

INTRODUCTION

The secret of getting ahead is getting started. The secret to getting started is breaking your complex overwhelming tasks into small manageable tasks and then starting on the first one.

— Mark Twain

In the last couple of days, I have been trying to reach the answer for a question that does not leave my mind. Does banking and financial business *really* have to be *that* complicated? Is the banking industry obsolete, and if so, why? As everything in life, my question began with a simple nuisance, around 15:30 on a regular Monday I tried to transfer money to my friend's account in another bank in the same country and he didn't get it until the next day. Appalling. Unreal. I shrugged and thought "Well, that's just the way it is". And it has been like that. 2020 has come to an end and we still have an old-fashioned course of action for the most everyday and basic things, like transferring money from one account to another.

I didn't want to use any of the third-party payment apps, like Revolut or Venmo, I wanted to spend and send my money tradition-ally and transparently. How could the infrastructure that should be evolving and progressing over the years fail on such a simple task? It beats me. I should know the answer to that riddle as I am head of ICT Channel in one of Croatia's biggest banks, but even though being at the root of the problem, I am still puzzled.

I asked myself, and millions of clients worldwide have been ask-ing themselves the same question. Why do we need to wait for the transaction in the digital age when we have smartphones, 4G, 5G, Blockchain, and many other gizmos and gadgets all around us? The clients don't care for heaps of paperwork, branch office appoint-ments, slowly awaiting death in lines for a meeting, coming again and again and again … to get a *simple* transaction.

DOI: 10.1201/9781003198178-1

I looked for an answer within the company, asked many experts in payment fields, and wasn't provided a meaningful answer. What I got is an explanation that payments are collected in one data file, which is being sent to a government Financial Agency every couple of hours, and the said Agency distributes items from XML files into appropriate banks – banks you want to send your money to. The last update that the Financial Agency gets is sent at 15:30 each day, every transaction performed after that will be executed the next day (Figure 1.1).

It sounds like something that was revolutionary when it was first invented, but why hasn't it progressed at all? To find the answers, I needed to dig in the past. The biggest reason is in the architecture of the banking information system that was built decades ago and most banks are still running their core functionality on it.

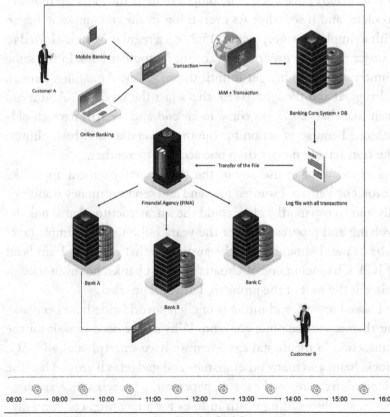

Figure 1.1   E2E payment process from Bank A to Bank B.

The Answer Is in the Core

You can think of the core banking system as the beating heart of any financial institution. Just like a heart, the core takes all the necessary services and business functions a bank provides and pumps it out to the bank's clients, because the bank and its clients make a complete body-like structure.

The core includes client management, risk planning, payment, security, card and credit processing, reporting, risk evaluation, position keeping, online-banking, mobile-banking interface, front offices, operational CRM, and much more. For processes to go smoothly, there are some things that need to be granted, such as robustness, availability, security management, and full audibility.

Infrastructure

Most of the banks are run on old and monolithic infrastructure, and to transform them into a new microservice architecture is a hard, time-consuming, and costly process. Monolithic architecture contains all core banking services in a single, fully integrated software section. A good thing is that you don't need to develop interfaces between services because, for example, something like the payment application and the position-keeping application is part of the same IT solution.

As the data model is consistent across all elements, there's no need to invest in complex features that connect different areas. The bad thing is that when you want to change something, you need to redeploy the entire system with performing a lot of regression testing (Figure 1.2).

Let's explain what is meant by "monolith" and "microservice". A monolithic architecture is built as one extensive system and is usually a single codebase. A monolith is often deployed all at once, both front, back, and the database with end code together, regardless of what was changed. However, a microservice architecture is where a software application is built as a chunk of small services, each with its individual codebase, spread on different infrastructure. These microservices are built around specific business capabilities and are usually separately deployable.

A monolith is built as a single, unified system. Usually, a monolith consists of a database, a front-end user interface (HTML or

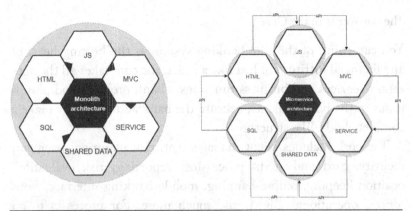

Figure 1.2   Monolithics vs. microservices.

JavaScript running in a browser), and a server-side application. The server-side application will handle requests, execute specific logic, retrieve, delete and update data from the database, and fill the HTML views to be sent to the browser.

In a monolith, server-side application logic, front-end client-side logic, background jobs, etc., are all defined in the same massive codebase.

Microservice architecture is nothing inherently "micro". Usually, they tend to be smaller than the average monolith; they do not have to be tiny. Some of them are, but the size is relative, and there's no standard unit of measure across organizations.

Microservice is an approach in which a single application is developed as a set of independent services, running in its own process and communicating with lightweight business processes, often with API. Microservices are built around business capabilities and independently deployable by fully automated deployment machinery using one of the cutting-edge tools, like UrbanCode Deploy or UrbanCode Release Management. There is a minimum of centralized management of these services.

Development

Why is development expensive? For a start, the highest cost (except infrastructure, data centres, servers, storage, memory units) is Engineer Labour and Engineer hours are expensive.

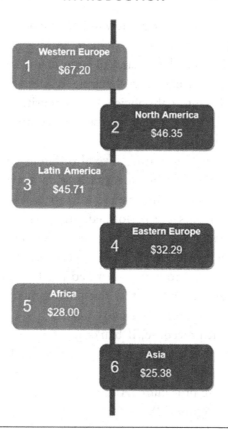

Figure 1.3 Average hourly rate of software development on 12/2020.

Take a look at the table above to conclude what is the average of Engineer hours across the world (Figure 1.3).

With these prices, you can easily calculate how much money you need to spend to "transform" any chunk of legacy software inside a brand new fancy software.

But the software development process is not about just development. For every hour of development that a bank needs to spend, you need to calculate on top of that 15% for project management, time for testing, and time for deployment.

Testing

Testing software applications in the banking domain is always challenging. Apart from common challenges banks are facing while testing in other fields, there are few unique challenges that are specific to the banking and finance domain. A huge effort is involved in testing activities for the enhancement or changes in the software applications due to frequently changing regulatory requirements, changes in bank's internal policies, highly sensitive customer data, and confidentiality requirements as well as complex systems integrated with each other.

Most of the banks are using several different environments for testing which include development testing on a separate test environment, component testing, where all test cases that cover new software requirements are performed. In most cases on this test environment teams are performing regression testing with connected components, modules, and applications. What to keep in mind is that the pre-production test bed needs to be a production replica with masked data. In this test environment teams are performing performance tests and smoke tests.

Following that procedure, if banks want to add a new field on GUI, it will cost at least five to seven Engineer Days + business analyst (at least a day for making a software requirement specification) for the entire procedure. And yes, it is expensive, and it lasts for a long time (Figure 1.4).

Let's circle back to the question: why is transfer of money expensive, and it can take from several hours to several days (yes, you need to pay to move money from one account to another), and why can't it be done immediately?

So, like any decent Generation Y member, I turned myself to the most popular search engine on the planet, and I found a term like "instant payment". What is an instant payment and when it will be integrated?

Instant Payment

Instant payments are initiated by the payer, i.e., the user of payment services, and their fundamental characteristics are the speed of execution, availability of use, availability of funds almost immediately (on weekends, at night, on holidays). Instant payments are becoming an alternative to cash or card payments and are available

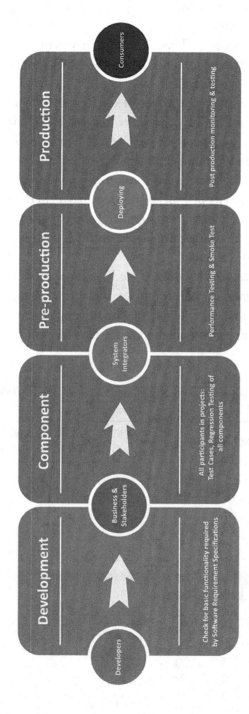

Figure 1.4 Life cycle of feature testing.

anytime, anywhere, and open up opportunities for all new innovative types of payments.

Within the European Union, instant payments for the Euro have already been in place since November 2017, under the name "SCTInst payment scheme".

Instant payments can be widespread. It is an intention for use in all segments and flows: consumers to consumers, consumers to business entities (payment for services at-home services, urgent payments, borrowing money), business entities to consumers (e.g. their employees), business entities to other business entities – suppliers (insurance payments, social transfers, payment of taxis, online trade, payment of obligations to suppliers) and payments to the state (payment of customs duties, fines, fees, taxes). It will also be suitable for small and large amounts.

Unfortunately, not all banks are ready to support SCT Inst functionality. In most European countries, instant payment will not be supported for years, and when banks finally implement the desired functionality, I am sure that the money transfer will be expensive (Table 1.1).

Table 1.1   SCT Inst Scheme Adherence Status Compared to the SCT Scheme Adherence in the EU as of Mid-June 2020

	SCT INST	SCT	PERCENTAGE OF SCT INST SCHEME PARTICIPANTS VS. SCT SCHEME PARTICIPANTS (%)
EURO ZONE			
AUSTRIA	452	502	90
BELGIUM	20	49	41
CYPRUS	1	17	6
ESTONIA	6	11	55
FINLAND	5	8	63
FRANCE	128	272	47
GERMANY	1,248	1,468	87
GREECE	0	23	0
IRELAND	2	205	1
ITALY	198	436	45
LATVIA	5	20	25
LITHUANIA	21	74	28
LUXEMBURG	3	65	5

(Continued)

Table 1.1   (Continued)

	SCT INST	SCT	PERCENTAGE OF SCT INST SCHEME PARTICIPANTS VS. SCT SCHEME PARTICIPANTS (%)
MALTA	5	31	16
NETHERLAND	9	38	24
PORTUGAL	16	36	44
SLOVAKIA	0	18	0
SLOVENIA	0	16	0
SPAIN	88	117	75
NON-EURO			
BULGARIA	3	24	13
CROATIA	0	21	0
CZECH REPUBLIC	0	21	0
DENMARK	1	53	2
HUNGARY	0	27	0
POLAND	1	27	4
ROMANIA	0	23	0
SWEDEN	2	9	22

While searching for instant payment options featured by banks, I ran into the latest report from Deloitte that has mentioned something about Banking Digital maturity at the end of 2020. I read it very carefully (you can do it too, by searching on the search engines: "Deloitte report 2020 Digital Banking Maturity"). I almost agreed with everything in the Deloitte report, but I couldn't see where 5G is in all that FinTech story. And it raised a question in my head: is it possible to build a bank but without the bank?

To answer that question, let's see first what will 5G bring to consumer banking.

"Is it possible to build a bank, but without the bank?"

5G Innovations

The newly coming of 5G raises the question of how mobile network operators can regain their investments. The market analysts expect expenses up to $1 trillion to roll out 5G networks by 2025.

Telecom operators have a seemingly long history of squandered mobile applications, financial services, and customer data opportunities. The arrival of 5G and IoT could provide mobile network operators an excellent chance to start a new innovation wave.

In retail and consumer banking, 5G will be mainly geared towards improving customer experience, building a new monetary ecosystem.

The legacy infrastructure can't provide connected devices with a digital wallet, process tenth of billions of transactions in real time through 5G applications nor offer competitive transaction fees.

What 5G Could Mean for Banks

In April 2021, 3.5 billion people worldwide use a smartphone. That's more than 50% of the entire world's population. It is no wonder that new technological advancements are in high demand these days and financial institutions are busy investing in technological upgrades.

According to a study conducted by AT & T and IDG, 81% of financial institutions have made some technology changes at the corporate and/or branch level in recent years (Figure 1.5).

Omnichannel is much more than just implementing multiple ways for clients to transact. It is a seamless and uniform interaction between clients and their financial institutions across various channels (mobile, online, ATMs, POS, branch offices). While multichannel is focused on transactions, omnichannel focuses on synergies.

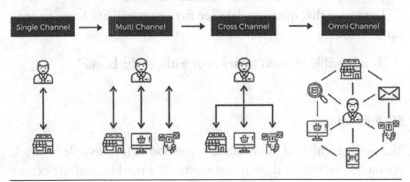

Figure 1.5 Multichannel vs. omnichannel approach.

A multichannel strategy allows access anytime, anywhere, on any device with a consistent experience across channels, while omnichannel enables interactions across multiple customer touchpoints where intents are apprehended, insights are derived, and communications are personalized and optimized. With omnichannel, banks can fulfil customers' precise needs and anticipate their wants and likes.

Now, 5G will for sure be a part of that technology upgrade. Telco companies like T-Mobile have already started implementing 5G network technology in some parts of Europe. In collaboration with regulations, like PSD2, Open Market approach and 5G will affect financial institutions and their customers.

5G networks have been tested in April of 2020 in Osijek, Croatia, and scheduled commercial work started in July of 2021. With this information in mind, I asked myself what will it do to society and what 5G eventually does for our financial institution?

It brings potential infrastructure upgrades and improves ATM connectivity. No longer wires will be needed to connect ATMs, which means greater location flexibility, they could be placed at the beach, forest, countryside, etc. It will provide faster internet and mobile access. Downloading a movie on the 4G network takes about six minutes, and on 5G, it will take only 15 seconds. This significant improvement allows much more information to flow through for financial institutions. Via 5G network, a bank adviser could walk a client through functions and stream instructional videos without downloading any banking application. It supports wearable payments, smart bracelets, watches, rings, necklaces, etc. 5G network will also help process payments through wearable devices. It allows access to a remote teller; this service will enable customers to get personalized attention with video sessions without visiting and waiting for the branch office appointment. Teller services will be available wherever the 5G network exists, whether on smartphones or via personalized ATMs.

Next-gen Customer Service

Banks will need to utilize this new technology when it becomes available. It is mandatory that they stay ahead of consumer expectations to ensure customers aren't tempted by a more agile rival.

With all the 5G network benefits, consumers will expect banks to instantly fix issues across all channels. Artificial intelligence and virtual reality will become table stakes for the customer offering, so banks need to decide how to leverage automated capabilities while still offering a personalized approach for clients.

Micro Banking Branches

We will see many micro banking devices in crowded places, because 5G technology can handle large traffic. With the help of 5G, the next-generation intelligent ATM-like devices will have highly integrated functions and use technologies such as artificial intelligence assisted self-service contactless interface with face recognition and digital signature.

We are still a few years away from 5G becoming mainstream in how 4G is today. It's useful to begin to think about how it might impact the financial services ecosystem. I focused on new experiences that clients can expect when connected to a 5G network with a 5G device in the list below. By 2022, we hope that 5G will drive accelerated mBanking growth, make video shopping experiences more widespread and compelling, make it possible for banks to deploy highly personalized customer service experiences, support time-sensitive banking applications, like online stock trading where milliseconds can judge between a win or loss, improve security, and fraud prevention. By computing and exchanging more data travelling between parties in real time, it will enhance mPOS transactions and utilization. 5G holds the potential to accelerate mobile point of sale transaction processing time and improve connectivity.

The future has already started. The banking industry needs to adjust, or it will disappear in the next decade. Modernization of the old monolithic banking system is a painful and expensive path but that path is needed; otherwise, new kids from the block will kill the 300+year old financial industry.

In the next chapters, we make futuristic architecture of the modern FinTech industry in which we don't need banks in the format like they are now. We don't want to "cut" banks from the process, we will transform them into Bank-as-a-Service.

In a Nutshell

What Will Be Good to Take from This Chapter?

$$\text{APY} = \left(1 + \frac{i_{\text{nom}}}{N}\right)^N - 1$$

APY = annual percentage yield
i_{nom} = nominal interest rate
(*N*) = number of compounding periods per year

Traditional Banking

- **Local branches are available** – Banks' actual buildings, called branches, that you can visit. These branches are staffed by tellers and other employees who can help you with all your banking needs.
- **Opening an account can take a while** – At a traditional bank, you will likely need to submit paperwork and visit a branch during normal business hours. You also have to wait to fully set up your account.
- **Some online banking options:** Traditional banks often offer a banking website or mobile app that you can use to complete your transactions. However, these apps are often not as robust as online banks.
- **Large ATM network:** Since traditional banks have bank branches, they also offer ATM access to their network of ATMs.
- **Lower interest rates:** You might get 0.10% APY (or even 0.01% APY) on a savings account at a traditional bank, but it's much easier to find higher yields at online banks.
- Typically, fees apply – a traditional bank might charge $10 or even $15 per month just to have a checking account.

- **Personal customer service**: Traditional banks definitely have an advantage here. After all, part of the cost they incur is staffing the branches with friendly employees.

Banking Core   A core banking system is a back-end system that processes daily banking transactions and posts updates to accounts and other financial data. Core banking systems typically include loan, credit, and deposit processing functions with interfaces to general ledger systems, business intelligence, and reporting tools.

Omnichannels   Omnichannel retailing is a multichannel sales approach that focuses on providing the customer with a seamless customer journey, whether the customer is shopping online, mobile, on the web, or in a real store.

Monolith Architecture   Monolithic software architecture is the traditional coupled model for designing a software program. Monolithic in this context means that everything is of one piece.

Monolithic system software is designed to be self-contained; the components of the program are interconnected and interdependent. For example, the front end is directly connected to a database. In a tightly coupled architecture, each component and its associated components must be present for the code to execute or compile.

In addition, if a program component needs to be updated, the entire application must be rewritten.

Microservices   Microservices – also known as microservice architecture – is an architectural style that structures an application as a set of services that are

- Independently deployable
- Highly maintainable and testable
- Loosely coupled
- Organized around business functions
- In the hands of a small team

The microservice architecture enables fast, frequent, and reliable deployment of large-scale applications. It also allows a company to update its technology stack.

Instant Payments Instant payment is exchanging money and payment processing that allow money to be transferred between bank accounts almost instantly, instead of the usual one to five business days.

2
TRADITIONAL BANKING

The greater danger for most of us lies not in setting our aim too high and falling short; but in setting our aim too low, and achieving our mark.

– Michelangelo

Beginning of the Banking

The banking systems have been around since the first coins were minted. Money or currency has been a part of human history for the last 3,000 years. I believe that the banking system has also been an integral part of society during human history, first by exchanging goods and later on by trading goods for gold (Figure 2.1).

In 1200 BC, human civilization found its way to counter the barter system's problems by introducing coins.

With social progression, the currency was also reinvented. Coins were made from gold and silver to standardize currency among the financial centres and enable trade. These financial centres were, in actuality, the temples and sanctuaries that were built by the older civilizations.

The historical records from Egypt, Rome, Greece, and Ancient Babylon depict the temples as banks, which kept the money safe and loaned it out to people.

Actually, the Romans were the first to initiate the banking systems. The relationship of lenders and borrowers was established after the takeover of Julius Caesar, who issued one of the edicts changing Roman Law. He allowed the bankers to confiscate the land or property of the person who was unable to return the due loan payments.

However, the Roman Empire eventually crumped, and likewise, the trade slumped, and the banks temporarily vanished. But some

DOI: 10.1201/9781003198178-2   **17**

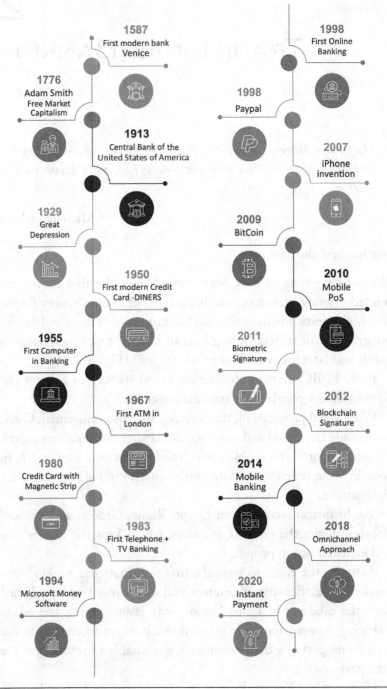

Figure 2.1   FinTech in time.

of those financial centres survived and emerged in the Holy Roman Empire in the form of papal bankers.

During the reign of monarchs over Europe, the banking institutions became stronger and stronger. Amid this period, the banks were formed under the power of ruling sovereignty. With all the power in kings' hands, they could easily get finance, which lead to unnecessary extravagances, an arms race with neighbouring kingdoms, and costly wars. Resultantly, the world's first national bankruptcy occurred in 1557, during the reign of Philip ll of Spain. His kingdom was burdened with too much debt due to pointless and unnecessary wars.

However, during the year 1776, the British empire was well established regarding its banking system when Adam Smith came along with his famous theory of "Invisible Hand".

In Adam Smith's perspective, the world's economy was self-regulated, which allowed the bankers and money lenders to limit the state's involvement in the banking system or the economy. And the concept of free-market capitalism and competitive banking was born, which was ultimately nurtured in the United States' soil.

Adam Smith's idea of the self-regulated economy did not benefit the American banks much, and the average life of an American bank was no longer than five years. The primary reason for the short-lived banking system was because most banknotes from the defaulted banks became worthless. Keeping this in view, a former Secretary of the treasury whose name is Alexander Hamilton established a national bank that accepted the banknotes at par and helped the banks float through difficult times.

The British were quite dominant in the banking sector in the late 19th century and up to the World War I. They also had considerable holdings in international finance, which they later shared with the German and French banks. These banks were actively involved in colonial and imperial finance.

Development of the Banking

In the medieval period, the first banks appeared. Most of these banks during the time were regarded as merchant banks. These banks' essential function was to lend loans and to finance the expeditions across the silk routes.

The most significant changes in the banking world took place somewhere between the 17th and 19th century. The Goldsmiths of London initiated the "proper" banks. Those banks revolved around the basic banking concepts like allowing deposits, issuing bank debt, etc. The first banks in England also started to give Promissory notes. Later on, the issuance of cheques and overdrafts began during the Industrial Revolution in the United Kingdom.

In the 20th century, the banking systems, a kind of which we know today, started to pop up. Post-World War ll events highlight that the banks began to lend money to countries, and retail banking came into existence.

In 1959, Magnetic Ink Character Recognition (MICR) code was introduced. It is a character-recognition technology used mainly by the banking industry to streamline the processing and clearance of cheques and other documents. MICR encoding is at the bottom of cheques and other vouchers and typically includes the document indicator, bank code, bank account number, cheque number, cheque amount (usually added after a cheque is presented for payment), and a control indicator. The format for the bank code and the bank account number is country-specific.

Scottish inventor Shepherd-Barron introduced the first Automated Teller Machine (ATM) on 27 June 1967. A machine was delivered into use in Enfield Town branch in North London by Barclays Bank in the United Kingdom.

On 3 May 1973, in Brussels, Belgium, The Society for Worldwide Interbank Financial Telecommunication (S.W.I.F.T.) payment network was established by Carl Reuterskiöld. Two hundred thirty-nine banks in 15 countries supported it.

SWIFT is a network which is used by banks and financial institutions to send and receive information, such as money transfer instructions. In 2019, more than 11,000 SWIFT member institutions sent approximately 33.6 million transactions per day through the network.

In the 1980s, the global banking system was laid down and regulated.

Computer ERA in Banking

The first core banking system appeared in the mid-1970s. These systems only provided the basic baking functionalities to their customers. In the last decade, banking architecture has evolved to provide several platforms to its customers, facilitating multichannel convergence. The digitalization of the world's IT systems also propelled the growth of online banking. The evolution of banking architecture is evident in the progression table below.

YEAR	EVOLUTION OF BANKING ARCHITECTURE
1970–1980	- Core basic banking systems were introduced. These systems only provided essential banking services like making transactions, reviewing account summaries, etc.
1980–1990	- Legacy core banking systems were introduced during this time. These were developed in silos and were primarily product-centric.
	- During this period, the Bank of Scotland initiated internet banking for the first time (TV + PhoneBox set).
1990–2000	- New core banking systems development. Those systems were customer-centric and more flexible.
	- In the mid-1990s introduced service-oriented architecture.
	- In the late 1990s multichannel integration and processing.
	- MS Money – personal finance management software program by Microsoft Corporation. It has capabilities for viewing bank account balances, creating budgets, and tracking expenses, among other features.
2000–2010	- The banking industry witnessed a huge increase in multichannel platforms.
	- Banks start to focus on customer-centricity, and online banking goes mainstream.
	- Core banking solutions were found through cloud-based platforms and data analytics (in the countries where the legislative allows it).
2010–2012	- Investments by banks into core architecture due to specific legal regulations (Data Protection Act, GDPR, etc.).
	- Rapid growth in mobile banking and focusing on risk management and security.
2012–2018	- Massive digital transformation by banks in their IT architecture. Development of new core banking solutions (monolithic architecture → microservice architecture).
	- Increase in online banking, payments, social networking due to customer inclination towards fast-growing digital services.
2018–	- Omnichannel approach → "Omnichannel" (also spelled omni-channel) is a consumer-focused approach that aims to provide a seamless shopping experience on all available (multiple) retail channels, including online websites, brick-and-mortar stores, mobile apps, TV, physical catalogs, radio, etc.

The Use of Computers and Internet in Banking

Society crosses into the digital world today, so banking without technology has become unthinkable. Computers have been part and parcel of the banking sector since the mid-1950s when the Bank of America introduced a computer that was designed for cheque processing.

Over the next couple of years, more groups began working on their machines. ATMs and online banking go side by side to facilitate each other's functions.

Each coming decade brought a new revolution to the world of computers and banking. With the introduction of ATMs in 1967, the banking sector changed the world. A week after John Shepherd-Barron debuted his first ATM in North London Barclays' Enfield Town branch, a Swedish cash machine was discovered; a month later, Westminster Bank in the United Kingdom rolled out its money machine. Over the next couple of years, more banking groups began working on their devices.

Online banking services were introduced in the year 2000 by the U.S. Banks, and within ten years, they successfully provided banking services to over 10 million customers. The new banking laws were introduced and enforced internationally.

In 2001, banks were considered the largest financial institution for supporting the state's industry and investment. In 2008, the financial crisis started because of deregulation in the financial industry that permitted banks to involve in hedge fund trading with derivatives. When the values of the derivatives crumbled, banks stopped lending to each other. The 2008 crisis was the most significant financial disaster since the 1929 Great Depression. It happened despite the efforts of the Federal Reserve and the U.S. Department of the Treasury. The crisis led to the Recession, where housing prices dropped more than the price fall during the Great Depression.

Introduction of Mobile

Mobile banking was not available till the year 1998. For the banking industry, mobile banking got a safe passage for use in the year

1999. Before online mobile banking, SMS or texts were primarily used to accomplish mobile banking, known as SMS Banking. For the first time, European banks initiated online mobile banking, which was supported via WAP support on the mobile sets.

Before 2010, Mobile Web Banking and SMS Banking were the two most popular products used to provide sleek customer care. The first mobile banking application service was introduced in 2010 to facilitate customers. With rapid development in the IT Industry and smartphone devices, mobile banking applications were introduced as well.

As these operating systems gained approval and smartphones became extremely popular among people, the banking applications were upgraded time by time. This improved customer banking and provided the clients with an enhanced user interface and financial transactional abilities.

To this date, all the financial institutions are using both mobile applications, push and SMS services, to transact their day-to-day operations. In addition to online transactions, the mobile application also provides clients with all other necessary information regarding their accounts, including alerts, updates, and other information. Nowadays, mobile banking services can be categorized into next five services: account information access, transaction carry-out, online investment analytics, support services, and content and news.

Introduction of Credit Cards

The first-ever credit cards were invented in 1950 when Frank McNamara and Ralph Schneider founded the Diners Club. They issued their first credit cards. The credit cards at that time were nothing as they are now. At that time, the credit card was actually a charge card that required its cardholder to repay the amount expended at the end of each month. Over the years, these credit cards evolved with the evolution of the banking system.

Along with the years credit cards with revolving credit limits were invented. With time magnetic strips were added, and then the

EMV chips were also introduced. A short chronological history of credit cards is described in the table below:

YEAR	TYPE OF CREDIT CARD
1950	Diners Club introduced the first credit card
1958	Bank of America offers a credit card with "Revolving Credit" for the first time
1958	Issuance of Entertainment and Travel Payment card by American Express Company
1969	Magnetic Strip Standards adopted in the United States
1976	Bank of America joins with other banks to create Visa and introduces BankAmericards
1979	Mastercard brand blooms into existence
1986	Discover card launched by the Dean Witter Financial Services Group
2015	To help buyers protect against fraud, EMV chips became a security standard

Besides the evolution of credit cards, Congress also passed many laws concerning credit cards. These laws were passed to protect consumer rights. A brief history of these laws is reflected in the table below:

YEAR	NATURE OF LAW
1968	Truth in Lending Act
1974	Fair Credit Billing Act
1977	Fair Debt Collection Practices Act
2009	Credit Card Accountability Responsibility and Disclosure Act

Monolithic Architecture of the Banking Software Systems

The core systems modernization stands for transforming or replacing critical applications while these systems need to work when the modernization is ongoing. Let me put this in a more plastic way; it is like performing brain surgery on the patient, while the patient is running a marathon.

Before going on about "brain surgery", we need to take one step back and declare that the banking core systems modernization is not legacy system integration; it is way more than that.

Firstly, legacy is a presumptuous term when applied to banking mainframe solutions. Legacy is defined as an outdated computer system. So, legacy is not a valid term in this context considering mainframe systems are IT's workhorses for more than 80% of the world's existing data, which are stored on mainframes.

This approximation varies by industry, COBOL and PL1 on CICS® (Customer Information Control System), IMS, and batch environments provide the infrastructure for the world's core banking applications. This approach will continue to be a base of banking IT architecture for the future.

CICS products are designed as middleware and support high-volume online transaction processing. A CICS *transaction* is a processing section initiated by an individual request that may affect one or more objects. This processing is usually interactive (screen-oriented), but it is also possible to process background transactions (Figure 2.2).

CICS Transaction Server lies at the head of CICS and provides services that extend or replace the operating system's functions. These services can be more efficient than generalized functional system services and more straightforward for a developer to use, particularly concerning communication with various terminal devices.

Although we witness an expansion in core banking solutions implemented on distributed systems, the built-in scalability, security, and reliability of the mainframe continue to be important determinants for the ongoing usage in Tier 1 and Tier 2 banks worldwide.

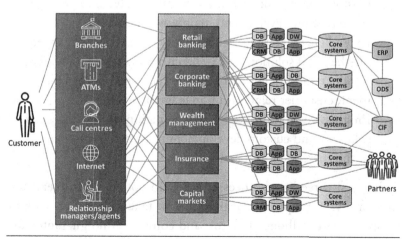

Figure 2.2 Monolithic core banking solution.

Most of the problems surfacing today in core systems result from tactical integration solutions with minimal consideration for the end-to-end solution architecture.

Current product-oriented core banking solutions fragmentation ends in increased maintenance. In 2005, banks reported that core systems account for one-half of total IT spending in a bank and estimate for three-quarters of ITs total maintenance. The non-flexibility within core systems limits the ability to respond to ongoing business requirements, which leads to constantly upgrading and repairing to keep the solutions relevant and viable.

To add a new financial product to existing CICS-based core systems required more than six months. As a result, banks are evaluating core systems alternatives to reduce costs and better align business and IT.

Modernizing Core Banking Systems

In most cases, there is no single suggested procedure. Business priorities, risk exposure, time to market needs are crucial dimensions that make a decision.

Core systems modernization falls into one of four primary areas:

- Package replacement: Picking and implementing a vendor package to substitute the current bank core systems.
- Custom development: Hiring an internal team of developers to design, develop, and maintain new systems.
- Extend current system functionality: Select and implement off-the-shelf solutions to extend current core function areas (product and pricing management, customer data management, customer relationship management, etc.).
- Progressive approaches: A combination of off-the-shelf solutions and custom development to extend (replace) parts of the current core systems.

The first three listed processes are valid options. Entire end-to-end core system replacement projects can take a couple of years to complete, cost millions of Euros, and come with no guarantee for success.

Progressive modernization extends (or replaces) core system functions incrementally, based on business requirements. With the formalization of microservice-oriented design approaches, the evolution of banking-specific master data management capabilities and industry models' maturation, this approach will be right as a roadmap.

Core systems modernization requires the linkage from business strategies with the chosen approach. Additionally, it is mandatory to have an active project/program sponsorship, usually at the C level.

Core system modernization with a business-oriented approach results in immense overall success. These projects are more successful in defining a complete business architecture (e.g. business models and associated processes, user groups, and data models).

After business segments are defined, solutions are designed through the definition of process models, business cases, and associated key performance indicators (KPIs). Those models and KPIs become critical inputs for software requirement specification or technical design.

A business-driven strategy enables banks to prioritize the domains for modernization based on market needs. Prioritization is made by phased development or initiatives for specific domains.

Most organizations pursue multiple workstreams together, but this multi-initiative approach is challenging; banks trying concurrent projects need to control the interdependencies between the projects and expectations. Disciplined range management is a crucial factor for achieving a good upgrade roadmap.

Solution design can be accelerated through the adoption of industry frameworks and reference models. Using models enables banks to customize best practices as part of the surgical process.

A well-defined business and technical architecture usually recommend a progressive modernization strategy. That iterative strategy consists of three major areas, is explained next.

Application upgrade focuses on applying a pattern approach for extending current mainframe core banking assets. It can be implemented in several ways, often in bottom-up development works. This method also provides for the integration of core solution assets with third-party solutions to enable the best strategies. Business process management (BPM) is a process-based method, usually starting from top-down modelling of business processes and use cases. And lastly, master data management for an account, customer, and data. The product application silos containing most core banking (mainframe) solutions lead to duplicated data stored in several different places for core bank data domains, such as customer and product data.

These strategies are non-exclusive; projects can use one or more of these above procedures as part of the roadmap. At the moment at UniCredit Group can be seen a more meaningful shift towards BPM and master data management-based approaches.

Progressive Modernization Architecture

The new solution's non-functional aspects must be consistent with the current conditions to meet the application consumers' expectations.

Banks are working with a large scale of internal users to assess their current environment and readiness for core systems renewal. The evaluation revealed product-currency gaps in their operational environments and monitoring deficits resulting in a recommendation to correct the infrastructure issues before embarking on a core systems renewal approach.

Non-functional requirements must be characterized during the modelling of every service, process, and data function. When evaluating approaches, several key non-functional areas should be assessed and evaluated, such as:

- **Performance and scalability:** The new solutions must maintain equal workloads with equivalent response times as the current system.
- **Availability and reliability:** Like performance and scalability, customers expect uptime and reliability to be consistent with the current system.
- **Security:** In this hacker age, ransomware, etc., security has added importance. Over the years, a vital attribute of the mainframe has been its rock-solid security at the hardware

and software level. By introducing the new solution components, new security considerations might occur and require the initiation of federated security solutions or new security components.

- **Management and monitoring:** By introduction, a new component solution, control, and monitoring requirements require re-evaluation. As a result, users need to assess system management resources to manage these solutions.

Banks must develop a coherent and logical path to a core systems upgrade. The current systems do not work; these challenged solutions block banks in many critical ways. Suppose we review the story around core systems modernization, bottom-up design, and line-of-business strategies that have driven siloed decisions, speeding these solutions decay. As we progress forward, it is clear that top-down design is necessary for successful core system upgrade projects. Without a robust business architecture and sponsorship, modernization projects are seriously challenged (Figure 2.3).

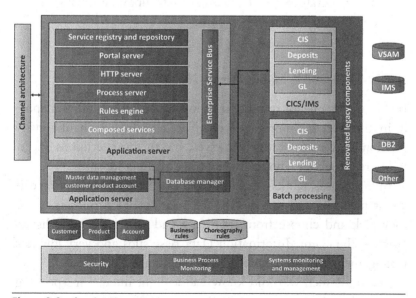

Figure 2.3 Core banking system.

Some of the lessons that we learned as we have engaged clients in core systems upgrade are:

- A good core systems modernization project benefits from a focus on one or two domains. Producing a roadmap based on market priorities, iterative improvements in core systems reduce the risk and better support evolving business requirements than full replacements.
- An industry model approach provides reference components to define the target business architecture and overall technical solution approach.
- Adopting a disciplined approach to service design, such as SOMA, is critical to developing business-relevant solutions and is key to application transformation.
- The improvement of a core systems modernization strategy links directly to the go-to-market process. By developing a roadmap based on market priorities, banks can precisely extend and replace their system's critical portions.
- ISV and custom approach for end-to-end core banking solutions. Success with core systems modernization results from thinking strategically and performing tactically, applying proven strategies to enable banks to upgrade during the day-to-day business operations.

Database Silos and Why They Aren't Good Anymore

Banks are famous for running in "silos". Loans in one office, deposits in different, and mortgages someplace else. In some banks, every division is left to fight for themselves, and turf wars erupt as uncertainty reigns about who really "controls" the relationship.

In most cases, banks often arrange the weight of contracting different products and services on the relationship administrator as they pick and choose from a product and services catalog list to point to the client. Information Data Base silos and very low risk management cost global banks billions (Figure 2.4).

The tradesmen who did know had a slight interest in speaking because they were earning so much money, often because the bank's

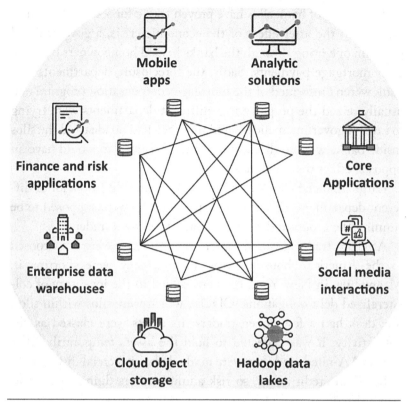

Figure 2.4   Siloed data environments.

accounting methods had miscategorized financial instruments in a way that concealed their risk.

In a large organization where specialization is vital, silos are natural. They can also be counterproductive; at Sony Corporation, three different groups worked on digital music players running three proprietary software to reinvent the Walkman player watched as Apple walked away with the reinvented music business.

In finance, the aim to create silos of functions and compensation has been reinforced by legacy information technology systems developed or purchased to meet individual financial department's needs. Modern systems are powerful enough to integrate processes across a bank, but managers often stop their adoption by managers who want to protect their turf.

The results of bank silos have proven tragic for some, like home-owners. In the after-effect of the economic crisis, a governmental program operating through the banks helped homeowners decrease their mortgage payments. Sadly, the foreclosure departments in a bank weren't informed of the mortgage compensation program and usually seized the property that different departments were trying to save. A government aide stated that they had no idea and the silos inside banks were so disorganized that the aid plan would have an opposite effect than intended.

Large financial companies and banks were split into a lot of different departments or silos; the managers who were supposed to be running the groups did not understand their own traders.

At UBS, traders assured senior managers they were not exposed to the American housing crisis; in fact, they were shorting it. Meanwhile, in New York, the bank added to the inventory of collateralized debt obligations (CDOs), which were silos within silos. The desk had a few dozen traders. Its assets were classed as client activity; it was expected to hold the assets temporarily. They were AAA-rated, and they were marked as commercial instruments rather than credit assets, so risk administrators didn't even know they existed.

The traders soon understood that given their almost free capital, even the 0.1% return gave them an impressive profit, which filled their bonuses. The danger, hidden by the grading system and the incentives provided by regulations, collapsed in August 2007 when the securitization market froze. Switzerland was astonished to learn that far from being short of the U.S. housing market, the New York holding would result in enormous losses, more than $30 billion eventually. At the bank, risk officers, who were divided into credit, market, and operational silos, never recognized the danger.

In a report after the losses emerged, UBS revealed it has several reports on real estate securities and loan exposure. However, there was no complete view available due to inadequate data capture, IT silos, in other words.

UBS improved its IT system and said it would never again suffer vital losses due to low-risk management. Later, it lost over $2 billion from unapproved trades by Kweku Adoboli, who was trading

ETFs, a problem that was not spotted because ETFs were relaxed in their small silo.

Organizations can overcome silos. The new organization's engineer hires are put through a six-week boot camp where the group stays together to learn best coding practices. The company also understood it could be a mechanism for social engineering. The engineers rotate through different fields so they can choose where they want to work. Every 12 months, engineers are asked to transfer to another project for a few months. Some stay in the new group, some return to the group where they had been working. It's unproductive in the short run, but it holds the organization fluid and busts silos and ends in information sharing among teams.

Security Point of View (from the Beginning to the Mobile Banking Era Short Overview)

The heist from the banks is steeping into mythology. But while the likes of Butch Cassidy got away with thousands of dollars in his time, today's bank criminals are more likely to conduct heists from a computer monitor.

When banks relied on police forces and simple alarms, the first bank robbery in history wasn't all that exciting. It was an internal job.

In 1798, Isaac Davis, an unnamed accomplice and a bank janitor, named Thomas Cunningham, took $162,821 from the Bank of Pennsylvania. Their method was straightforward: Cunningham slept at the bank. Their wealth was short-lived. Davis admitted to his crimes after depositing the stolen money into the bank he robbed.

As development headed west, the region saw a growth in bank robberies. Two men became famous at this time: Jesse and Frank James. In a robbery of the Clay County Savings Association in Liberty the two went away with $60,000.

As the 19th century came on, Butch Cassidy and Sundance Kid made the classic bank robber's model. In 1889, the group took $20,000 from Miguel Valley Bank in Telluride. Their method was easy, too. Dressed in expensive clothes, Cassidy occupied a bank worker before putting a gun to his head. He later escaped using a range of safe houses.

In the early 1900s, Cassidy's Wild Bunch specialized in a different sort of American trope: the train burglary. The high-profile scene was part of the mind the group grew to such a notorious reputation.

Bonnie and Clyde's doomed romance narrative stole headlines; John Dillinger grabbed a pile of money in the 1920s and the 1930s, which was something of a bank crime renaissance.

Dillinger's practices were very creative. His organization once posed as a film crew executing a bank heist for a movie. The second time, he passed himself off as a bank alarm salesperson. His group was responsible for countless robberies between 1933 and 1934. Despite their fame, they were particularly violent, with Dillinger's assistant Baby Face Nelson prized for gunning down a police officer. The group made more than $300,000 from bank robberies.

The Boyd Gang was responsible for countless bank robberies. They defeated security measures like high-walled booths and metal bars to get away with what today would be millions of dollars.

By the 1980s, half of all bank robbers still flashed a gun in the process of taking money. But as security technology evolved, robbers had much more challenges in robbing a bank.

Exploding dye packs and bait money, silent alarms covertly contacted police enforcement to help keep cashiers safe, CCTVs captured in-person heists on camera for posterity.

These measures have shown promising results. For example, in Boyd's Canadian home turf, 70% of the robbery was solved in the 1990s. By the 2010s, that figure was around 90%. In addition to securing money, these practices helped reduce violence in robberies.

In the 1990s, we witnessed a peak in bank robberies in the United States, with around 9,000 in a year, compared to 3,500 in 1975. But the way those robberies went down had changed dramatically from the days of stagecoach holdups.

By 2000, only a third of bank robbers took firearms as they made out with cash. In the last couple of years, that number has dropped to about a quarter. One reason for that drop is because bank tellers are known to be unarmed.

Today, a paper note is sufficient for a robber to get away with cash. 2,416 bank heists occurred in 2015 with a demand paper note. In the same year, firearms were used in only 877 bank robberies. Today, Butch Cassidy is more likely to be a hacker. With many banks moving online and leaving branches, stealing money online makes more sense.

Cyberattacks are also more skilled at stealing money. One group went off with $101 million from a bank in Bangladesh in 2016. In terms of security for general digital transactions, banks use antivirus protection programs, firewalls, and encryption to support secure and private online banking logins from customers.

The biggest threat today to your money is your online behaviour. With consumers frequently making purchases and banking transactions online, hackers find innovative ways to take your identity, credit card information, and bank logins.

Even if the bank is hit with a robbery, fraud, or even a natural disaster, customers aren't likely to encounter a loss. And, in the case you have fraudulent attacks on your account or unauthorized money transfers, your bank will likely issue you a credit.

In the end, cyberattacks are more prominent than ever, and bank robbers are less violent and less of a risk to clients. With added insurance and securities, banking customers aren't likely to have to foot the bill when a modern-day Sundance Kid gets away with cold cash.

In a Nutshell

What Will Be Good to Take from This Chapter?

Modernization of the Core System Modernization of core systems represents the conversion or replacement of critical applications, and those systems must continue to function during the modernization. It's like performing brain surgery on a patient while he's running a marathon.

Before we talk about "brain surgery", we need to take a step back and explain that core banking system modernization is not an integration of legacy systems; it is much more than that.

First, "legacy" is a lofty term when applied to mainframe solutions in the banking industry. Legacy is defined as an obsolete computer system. So legacy is not a valid term in this context, considering that mainframe systems are IT's workhorses for more than 80% of the world's existing data stored on mainframes.

Modernizing core systems requires linking business strategies to the chosen approach. Additionally, active project/program sponsorship, usually at the C-level, is mandatory.

Core system modernization with a business-oriented approach leads to immense overall success. These projects are more successful in defining a complete business architecture (e.g., business models and associated processes, user groups, and data models).

Database Silos Data silos are an accumulation of data held by one group that is not easily or fully accessible to other groups. Internal departments usually organize the data. Finance, administrative, human resources, and other departments need different information to do their jobs, and these individual collections of often overlapping but inconsistent data reside in separate silos. As the volume and variety of data continue to grow, so do the silos.

Data silos hinder the process of gaining deep, actionable insights from enterprise data and are a barrier to a holistic view of enterprise data. To get the full benefit of data analytics, companies need a 360-degree view of their data to gain an enterprise-wide view of hidden opportunities.

3

BLOCKCHAIN

The person who doesn't know where his next dollar is coming from usually doesn't know where his last dollar went.

– Unknown

What Blockchain Means to the FinTech Industry?

The digital revolution has changed the banking and FinTech industry. This industry started to undergo a total digital transformation to make services more secure and consumer-friendly. Now when financial transactions go through digital channels, new cyberthreats cost businesses customers and revenue. The new Blockchain approach can secure a full stack of security in digital commercial transactions (Figure 3.1).

Blockchain technology is bringing a new form in accomplishing tasks with ease and reducing risk.

Paper currencies will become obsolete with digital transactions driven by the 3G, 4G, and 5G mobile revolution. A step further into digital-based transactions are today's cryptocurrencies, an alternative form of digital currency.

Unlike conventional digital currency that banks are using, cryptocurrencies work on "decentralized" control that uses a distributed ledger technology called "blockchain". The Blockchain is fast-emerging as a public financial transaction database for the secured digital transaction, including cryptocurrency.

A ledger is a book containing accounts, and summarized information from the records is posted as debits and credits. The ledger contains information that is required to prepare monetary-financial statements. It includes accounts for assets, liabilities, owners' equity, revenues, and expenses.

DOI: 10.1201/9781003198178-3

Figure 3.1   Blockchain overview.

How Do Blockchain Transform Banking and FinTech Industry?

Banking is a crucial area of the financial sector that is more sensitive and vulnerable to cyberthreats because of the vast reserves they hold in their databases. In recent years, various banks worldwide have reported severe cyberattacks that involved a direct strike on centralized databases, creating the loss of countless billions of dollars.

This eventually resulted in governments issuing strict directives, and now, large banks have started looking at ways to adopt advanced decentralized asset solutions like Blockchain.

Surveys show that the top managers of global banks are starting to use Blockchain in the financial sector due to technology's ability to decrease costs and add security.

Blockchain gives a considerable level of security when receiving and transmitting data. It secures an open and transparent network infrastructure while allowing a decentralized and low-cost operation approach. All this Blockchain allows being an attractive solution for businesses in the financial sector.

Current transactions involve mediators to facilitate transactions, which makes the end-user banking services cost more. Blockchain holds a unique advantage by bypassing the need for an additional medium to perform transactions and provide the services at a lower cost.

Some of the critical advantages Blockchain offers over traditional payment systems include:

- Faster bank-to-bank and foreign transactions at a lower cost
- A single customer identification system that stores client details in a single instance

- Shares customer information with other banks in a secure way
- Blockchain can substitute SWIFT transfers

Global Banks that have already tested and implemented Blockchain reports can meet legal requirements and comply with data protection regulations while using Blockchain solutions.

Blockchain is also considered an effective method of data processing and storage and credibly performing authorized transactions.

With further optimized payment facilitation, the distributed ledger system of Blockchain reportedly accelerated transaction speeds while also saving costs associated with processing transactions.

Many companies have also reaped Blockchain's benefits for many other financial market utility services involving clearance, settlement, and other intermediary functions.

Blockchain comes as a safe and efficient method of dealing with digital transactions, securing efficiency in processing transactions of any size in a quicker time frame, and enhanced cost-savings. This ultimately results in enhanced quality of end-user services with efficiency, security, speed, and cost savings.

What Exactly Is Blockchain in Relation to the Commercial and Banking Industry?

Blockchain is an open, distributed ledger that records transactions between two parties efficiently. A Blockchain is made of blocks of data that involve a series of linked transactions, joined together in a specific order.

All involved parties can share a digital ledger across an internet network without requiring a centralized authority or intermediaries. That's the reason why processing transactions within Blockchain is faster.

The speed is just one of the benefits that Blockchain brings to the financial and banking industry. It's not only about more comprehensive efficiency. It also gives a new level of transparency and security.

Behind the Curtain Mechanics of the Blockchain

When Frank decides to transfer a bitcoin to Nancy and clicks send, Frank's wallet should have one less bitcoin – and Nancy's should have one more.

At this point, the verification process begins. The transaction request from Frank is broadcast to the entire network. Anyone on the network can use the public key to confirm that the transaction request came from the legitimate account owner. But the transaction is not yet verified. This is where miners come in (Figure 3.2).

The verification process ensures that Frank can send Nancy bitcoin. All digital currencies have Blockchains with their unique mechanics, but in the case of bitcoin, a new block is added to the

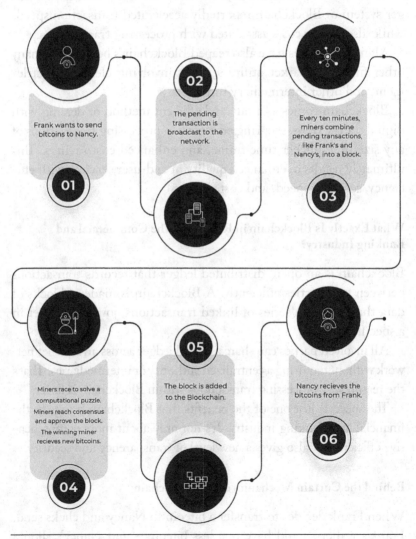

01 Frank wants to send bitcoins to Nancy.

02 The pending transaction is broadcast to the network.

03 Every ten minutes, miners combine pending transactions, like Frank's and Nancy's, into a block.

04 Miners race to solve a computational puzzle. Miners reach consensus and approve the block. The winning miner recieves new bitcoins.

05 The block is added to the Blockchain.

06 Nancy recieves the bitcoins from Frank.

Figure 3.2   Blockchain transaction.

Blockchain every ten minutes. Miners, the specialized computers on the network, compete to pack the data from Frank and Nancy's pending transaction with other unrecorded transactions into a new block (let's assume that the block containing Frank and Nancy's transaction is block #100000). The previous block (#99999) is included in the miners' process and a random number known as a nonce. Miners compete to solve mathematical calculations associated with block #100000 to win a prize: newly generated bitcoins.

Transactions are verifiable when the miners create a unique cryptographic fingerprint using a hash function. The hashed block must have a random number of zeroes at the beginning (Figure 3.3).

A hash function takes a set of digital data and returns a numeric piece of data with a fixed range. If you pass exactly the same data to a hash function, it will return exactly the same numeric data every time. If the data input varies by even one variable, the output of the hash function changes.

The hash with the right number of zeroes is completely unpredictable, so miners keep trying different hashes. When the winning miner finds a hash with the correct number of zeroes, the discovery is announced to the rest of the network. The other miners acknowledge the recognition and immediately turn to the next block (#100001), whose element is the newly verified block #100000. The Blockchain code then rewards the miner who verified block #100000 with 12.5 BTC. The hashed block gets a timestamp and is published, which means that block #100000 is added to the chain of existing blocks. The Blockchain goes back to the first block, which is called the Genesis block (Figure 3.4).

Receiving the Digital Currency

All that may sound not very easy, but for Frank and Nancy, it is relatively straightforward. Frank sends the bitcoin to Nancy via his digital wallet, and shortly after that, it arrives in Nancy's. A Blockchain network removes intermediaries and simplifies the process.

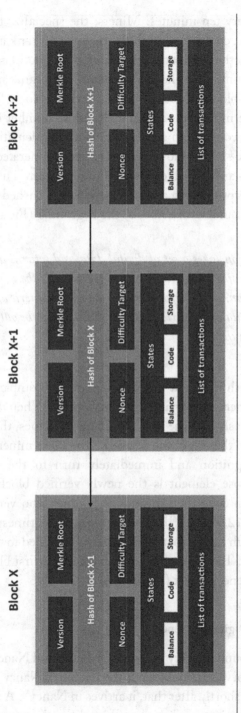

Figure 3.3   Blocks in a Blockchain.

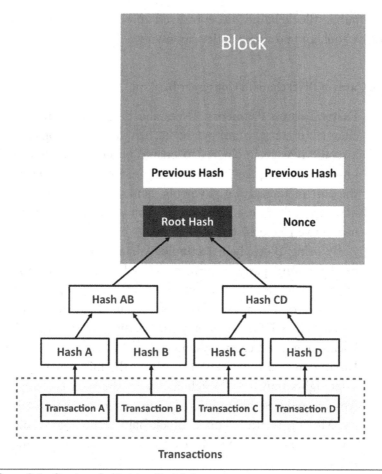

Figure 3.4 Structure of a block in a Blockchain.

How Can Banks and FinTech Benefit from Blockchain Technology?

Once we consider the benefits of Blockchain, it only makes sense that the FinTech and banking industry will take the lead in the adoption of it.

Why humanity invented banks in the first place? Financial and banking institutions were created to unite groups of people and allow trade and commerce among them. A Blockchain is a secure and transparent tool that can accomplish the same but on a global scale.

Blockchain holds the potential for global commerce. Removing the manual processes and introducing streamlined and automated one's Blockchain could make trade more efficient.

A public Blockchain is decentralized and can be a great collaborative tool, and no single entity can own it.

Use Cases of Blockchain in Consumer Banking

1. **Faster/Instant Payments.** By establishing a decentralized channel (e.g. crypto) for payments, banking institutions can leverage new technologies to enable faster payments and lower processing fees. By increasing security and lowering the cost of transaction payments, banks could introduce a new level of service, offer new products, and compete with innovative FinTech start-ups.

   By using Blockchain technology, banks can reduce the need for third-party confirmations and speed up processing times for traditional bank transfers. In 2016, 90% of members of European Payments Council believed that Blockchain will fundamentally change the industry by 2025, Clearing and settlement systems.

2. **Buying and selling assets.** By cutting the middleman and asset rights transfer, Blockchain lowers the asset exchange fees and reduces the traditional securities market's instability. Moving securities on a Blockchain technology could save up to $24 million each year in foreign trade processing costs.

   Buying and selling assets are based on keeping track of who owns what. Financial markets achieve this through a complicated system of exchanges, brokers, clearing houses, central security depositories, and custodian banks. All of these various parties have been constructed around an obsolete method of paper ownership.

   Executing transactions electronically is complicated because buyers and sellers don't use the same custodian banks, and these banks don't always rely on third parties to hold onto all the certificates.

   The system is not only disorganized, it is inaccurate. Blockchain will transform financial markets by creating a decentralized database of digital assets. A distributed ledger allows the transfer of rights to an asset through cryptographic tokens that represent those assets outside the chain. Cutting out the middleman will reduce fees and speed up the process significantly.

3. **Loans and credits.** Traditional financial institutions underwrite loans using a system of credit reports. With Blockchain, we see the future of peer-to-peer lending, faster and more secure loan processes, and even complex programmed loans that can approach the structure of syndicated loans or mortgages.

Banks that process loan applications assess risk based on factors such as credit score, homeownership status, or debt-to-income ratio. Such centralized systems are often harmful to consumers because they contain faulty information. Moreover, the concentration of sensitive data among a small number of institutions makes them highly vulnerable.

4. **Digital verification of identity.** Banks wouldn't be able to process financial transactions without an Identity Access Management process. The IAM process consists of many different steps that clients don't like, Face-to-face checking, a form of authentication, biometric authentication or authorization. For safety reasons, all of these steps need to be repeatedly taken for every new service provider.

With Blockchain, clients and companies will benefit from accelerated Identity Access Management processes. Blockchain will make it possible to reuse identity confirmation for additional services securely.

The most popular innovation in this domain is Zero Knowledge Proof. Several corporations are now working on solutions based on ZKP. Thanks to Blockchain, users will choose how they wish to identify themselves and with whom they agree to share their digital identity. Consumers will need to register their identity on the Blockchain only once. There's no need for repeating that registration for every service provider – as long as the Blockchain powers those providers.

The Future of Blockchain in Banking

To make the most of Blockchain, banks first need to upgrade the current infrastructure required to operate a global network using matching solutions. Only a general adoption of Blockchain will lead this technology to disrupt the financial sector.

Although the investment in modernizing monolith architecture will come with significant returns, Blockchain will enable banking institutions to process payments faster and more precisely once fully adopted.

All in all, blockchain-enabled financial and banking applications will deliver a better customer experience that can compete with FinTech start-ups.

Know Your Client (KYC), the regulatory process used to verify clients' profiles, is also set to impact the industry significantly. This regulation makes attracting new customers increasingly demanding and complicated for the bank and customers, who have to complete endless forms.

Some KYC information may be factored in and pooled between banks. "Regtech" start-up companies are already offering products of this character, and customers have already shown interest in such solutions.

Blockchain shares and gives information and encryption in ways that make it a compelling solution for KYC obligations. Blockchain can also be deployed on private networks, so cryptography can be implemented to shield delicate information.

It's almost impossible to create a distributed KYC service without sharing some delicate information. There is a possible viable business model for scenarios when a client performs a KYC process with Bank A before deciding to do financial business with Bank B. In this case, the client could use a Bank A certificate to prove to Bank B that he has already finished a KYC process. Bank B would probably pay compensation to Bank A, but this would be insignificant compared to the cost reduction.

In this model, Bank A would be informed that its client has started a relationship with Bank B. When multi-banking is common, at least in the private banking world, this would not be an issue.

Payments Revolution

Blockchain is also a mechanism that can facilitate cross-border transactions. For example, a cryptocurrency, Ripple, has set up a Blockchain network uniting banks. The everyday use case is a cross-border transaction between two SMEs. The regular correspondent banking scheme is relatively ineffective; it takes five days to shift the money, while transaction fees soon add up. There is also a risk that the intermediary bank will default with the standard scheme, leaving transactions unmatched.

Blockchain offers a more cost-effective solution. In the case of Ripple, an internal coin has been created. Payment is done in minutes and doesn't depend on message correspondence. As the internal currency is not an actual digital currency, each bank still needs money to stake the transactions. Despite this minor drawback, this solution is up-and-coming.

The PSD/2 directive also disrupts payments in Europe. Among other things, this enables clients to manage all their accounts with different banks using only one interface. This revolution on the front side massively impacts the banking landscape. A similar process in the back office could accompany it. Blockchain technology would introduce payment systems that work using transactions instead of messages, making a clearing mechanism unnecessary.

Beyond Payments

Blockchain technology can be used to make smart agreements or smart contracts. The advantage is smart contracts are executed automatically according to predetermined criteria.

In this instance, Blockchain can monitor financial products' life cycle, which is more complex than simple payments, including options and other derivative instruments. Due to their derivative nature, whether they are OTC or listed, these products demand both parties to check market conditions and track when barriers are reached.

Cash flows are generated when certain market conditions are reached. Following several hundreds or thousands of products consume significant resources and demand substantial reconciliation work. It would be more useful to automate the process using smart contracts to schedule these cash flows.

Speed of Adoption

The speed at which Blockchain will be adopted will be determined by differences in the structure of the financial market from country to country. The development will be faster in more modest or non-existent ecosystems. Australia could adopt Blockchain more quickly than other countries as there are only four banks. The same is in Switzerland, where integration is strong thanks to the SIX centralized payment system.

Security will also benefit from this technology. Adopting it could take between 10 and 15 years due to the lack of ambition to transform the financial industry.

In the coming months, we can expect to see the first Blockchain initiatives reach the production stage. These challenges will not necessarily be more complicated than those we already face in the area of security (Figure 3.5).

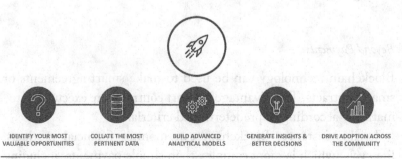

| IDENTIFY YOUR MOST VALUABLE OPPORTUNITIES | COLLATE THE MOST PERTINENT DATA | BUILD ADVANCED ANALYTICAL MODELS | GENERATE INSIGHTS & BETTER DECISIONS | DRIVE ADOPTION ACROSS THE COMMUNITY |

Figure 3.5   Blockchain solution.

Despite the lack of standardization of Blockchain solutions, this technology is here to stay; it allows all assets to be truly digitized.

In a Nutshell

What Will Be Good to Take from This Chapter?

Blocks Blocks are files that permanently record data pertaining to the Blockchain network. A block records all of the most recent crypto transactions not yet included in previous blocks. Therefore, it is like a page of a ledger or a record book.

Ledger A distributed ledger is a shared database and synchronized across multiple locations, institutions, or geographies and to which numerous people have access. It allows transactions to have public "witnesses". The participant at each node in the network can access the shared records across the network and can have an identical copy of them. Any changes or additions made to the ledger are reflected and copied to all participants within seconds or minutes.

A distributed ledger is in contrast to a centralized ledger that most companies use. A centralized ledger is more vulnerable to cyberattacks and fraud because it has a single point of failure.

The technology underlying distributed ledgers is the same as Blockchain, the same technology used by bitcoin.

Hash Code A crypto hash function is a mathematical function that converts an input of unpredictable length into an encrypted output of fixed length. Therefore, the unique hash is always the same size regardless of the original amount of data or file size. Also, hashes cannot be used to reverse engineer the input from the hashed output because hash functions are "one-way".

However, if you apply such a function to the same data, its hash will be identical; therefore, you can verify that the data is the same if you already know the hash. Hashing is also essential for Blockchain management of cryptocurrencies.

Mining   Blockchain mining is a peer-to-peer computer process used to secure and verify bitcoin transactions. Blockchain miners are involved in mining by adding bitcoin transaction data to bitcoin's global public ledger of past transactions. Blocks are secured in the ledgers by Blockchain miners and linked together to form a chain.

If we go in depth, unlike traditional financial services systems, bitcoins do not have a central clearinghouse. Bitcoin transactions are usually verified in decentralized clearing systems where people contribute computing resources to verify the same.

This process of validating transactions is called mining. It is probably called mining because it is analogous to mining materials like gold – mining gold demands a lot of effort and resources, but there is a finite supply of gold; consequently, the amount of gold mined each year stays about the same. In the same way, mining bitcoins consumes a lot of computing power.

4

5G Opportunity

It's not the employer who pays the wages. Employers only handle the money. It's the customer who pays the wages.

– Henry Ford

5G Opportunity

In the banking and financial industry, 5G will be organized towards improving customer experience. The roll out of 5G with speeds far better than 4G will enhance customer experience and allow for real-time processing, making connections more secure.

5G technology will revolutionize multiple sectors of society. This technology speeds ten times faster than 4G, will also make possible higher connection density (increased number of connected devices for each receiving unit) and decreased latency. The most significant impact will be manifested in self-driving cars, virtual reality (VR), smart cities, remote surgery, in the financial sector and banking industry.

Pilot projects in the financial sector will consist of access to monetary market services that require strong communication with very low latency. Experts agree that 5G will be mainly geared towards improving customer experience in banking and financial services. It means a new way of delivering products and services to the end clients. It will be normal to purchase products that had previously been presented in virtual or augmented reality (AR).

5G will make better services available to the end clients so financial transactions will take place anytime, anywhere. Virtual assistants will enhance basic online transactions in real time, enabling more complex queries that are currently not possible.

DOI: 10.1201/9781003198178-4

5G will improve banking applications and infrastructure. Core banking applications and infrastructure will have greater capacity, and therefore more powerful information analysis with higher volumes of data will be possible.

The arrival of 5G is followed by what is known as "edge computing". It will reduce network latency, bringing services closer to users. Network nodes will have a standalone computational capacity, and information will not have to go to the cloud to be processed. Data will not need to travel long distances, cloud workload will be reduced, and real-time analysis will be engaged.

Also, improving customer experience and bank infrastructure, processing information in real time will make it possible to improve communications security.

Banking security software tools can be improved with more precise biometric technologies with the enhanced power of processing. And these confirmations can take place on a remote server, given the low latency and high speed of the network.

Whereas 4G networks do not have any native prioritization of services, one of 5G networks' best features is the so-called "network slicing".

Network slicing is the ability to segment the network, offering priority, and confirmed services. We also need to note that any new technology comes with security questions, with its characteristic technologies like "slicing", making transactions and financial operations more secure.

Edge Computing

Edge computing is remodelling how data is being handled, processed, and delivered from devices worldwide. The explosive growth of the IoT devices and new applications that require real-time computing power continues to drive edge-computing systems.

Early goals of edge computing were to lower bandwidth costs for data that needs to travel long distances. The rise of real-time applications and, as we mentioned, an explosion of IoT devices that require real-time processing will drive the technology further (Figure 4.1).

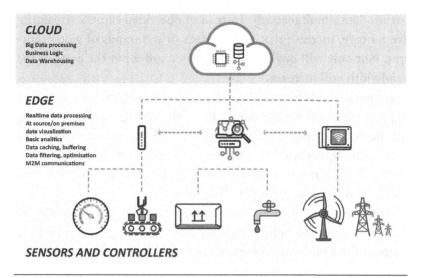

Figure 4.1 Edge computing.

What Is Edge Computing?

It was developed because of the exponential growth of IoT devices that connect to the internet to receive information from the cloud or transmit data back to the cloud. And many IoT devices produce huge amounts of data during their operation.

Gartner defines edge computing as *"part of a distributed computing topology where information processing is near the edge – where things and people produce or consume that information"*.

At the most basic level, edge computing brings computation and data storage closer to the devices where it is collected, rather than sending it to a central location that may be thousands of miles away.

This eliminates latency issues with real-time data that can affect an application's performance. Companies can also save money by having the processing done locally, reducing the amount of data that needs to be processed in a cloud-based location or an on-premise system.

A great example of using edge computing can be an internet-connected video camera streaming live footage from a remote location.

While a single camera producing data can transmit it across a network very easily, problems occur when a large number of devices stream data simultaneously. Instead of one video camera streaming live footage, increase that by hundreds or thousands of video cameras. Not only will quality due to latency suffer but the costs of the bandwidth will increase.

Edge-computing solves this challenge by enabling a local source of processing and storage for many of these gadgets. For example, IoT devices can process data from that IoT device and then send only the relevant data back through the internet on a cloud, decreasing bandwidth needs or sending data back to the IoT device in the case of real-time requirements (Figure 4.2).

For many organizations, cost savings may be a reason to implement an edge computing architecture. Organizations that have adopted the cloud for their applications may have found that bandwidth costs were higher than expected.

The most significant advantage of edge computing is the ability to process and store data faster, which allows for more cost-effective processing of real-time applications that are critical to businesses.

In an edge computing paradigm, the algorithm runs locally on a server or even on a smartphone device. Applications such as virtual and AR, smart cities, self-driving cars, and even building automation systems require fast processing and response.

Smart Banking in 5G World

Current bank branches are aiming to transform into the next generation of scenario-based units. AI + 5G + cloud will make more high-quality bank branch applications a reality, giving powerful support for smart bank branches.

With the FinTech and banking industry development, internet finance, mobile banking services, and self-service have become more popular.

Current bank branches are aiming to transform into the next generation of scenario-based units. AI + 5G + cloud will make more high-quality bank branch applications a reality, giving powerful support for smart bank branches.

Most of the banking services can now be processed on mobile devices. Given this development, traditional bank branches see a

Intelligent building
275 GB per day (0.1% transmitted)

Smart Hospital
5 TB per day (0.1% transmitted)

Smart Car
70 GB per day (0.1% transmitted)

Smart Grid
5 BG per day (1% transmitted)

A city of
one million
will generate
200 milion GB
of data
per day!

Connected Plane
40 TB per day (o.1% transmitted)

Public Safety
50 PB per day (<0.1% transmitted)

Connected Factory
1 PB per day (o.2% transmitted)

Weather Sensors
10 MB per day (5% transmitted)

Figure 4.2 Where are big data generated?

fall in customer traffic. Also, customer acquisition ability is similarly declining.

With the internet and smartphones' availability, making access everywhere, more personalized mobile financial services – offering a superior user experience – are emerging.

We can be misled and conclude that this will lead to bank branches becoming obsolete, that is not the case. The development of artificial intelligence and 5G will empower bank outlets by allowing them to become intelligent. Branches will still play an essential role in the smart era but with new functions and objectives.

Despite the rapid development of mobile services, Omnichannel, and self-service, banking branches remain integral in high-value fields, wealth management, consultancy, and private banking. In these scenarios, clients like face-to-face communication.

Branches remain especially important for brand promotion. Only in branches clients get an all-around experience. Bank branches can combine online (via Kontakt Centar) and offline activities to promote service development.

Smart branches will involve multiple intelligent, self-service, and remote video devices to provide promotion and training services. With improving customer experience, labour costs for banks are reduced. Branches are evolving to become social experience centres that are scenario-based. They are blending with other industries as entrusted platforms that give business and financial services.

These smart branches will be lightweight. Limited physical space will be used effectively. Operations and Maintenance (O&M) will be very efficient. With this merging of online and offline functions, the ability to gain new clients will inevitably improve.

A last but essential, fundamental feature of smart branches will be mobility. 5G wireless connections will provide well-built and large data services, giving widespread financial and business services coverage.

Smart branches will be reconstructed to lead customers to smart equipment to handle services, and fewer customers will be led to the counter. Service personnel is available to offer guidance to clients at any time they need it. To reduce waiting time, online services will deliver a brand new experience.

New technologies (AI, 5G) can enhance a bank's network design, provide a full-journey network experience for clients, and optimize network control methods.

When a client steps into a smart bank branch, service personnel can immediately identify the individual customer using a smart device. An interactive screen displays products that the client has browsed on their application (Omnichannel approach).

Through VR and AR, the client will not need to go to the main branch office to handle complex business services.

The service personnel will contact experts who are ready to answer client questions in real time. Elsewhere, the temperature, humidity, and even the building's lighting can be automatically adjusted according to the people's density in the branch office and weather conditions.

The Architecture of Smart Bank Branch

The smart solution provides a complete set of customer services for bank branches. Four banking scenarios are critical in intelligent branches:

1. **Smart connections:** With the rise of intelligent branches, applications, and high-definition video, high bandwidth capacity becomes essential. The network quality of branches is crucial. It is necessary to use 5G technology to give large-bandwidth network coverage and enable intelligent selection of internet lines, ensuring maximal bandwidth for bank key services.

2. **Smart marketing:** AI, big data, and cloud platforms will be used to build an online and offline precision marketing platform (Omnichannel approach). With AI capabilities, intelligent customer services will guide customers in branches. Smart cameras (CCTV) can accurately identify VIP bank clients on arrival. Big data analytics of the precision cloud marketing platform implement personalized service support and highly customized marketing.

Also, TV displays in the branch office will be used to show marketing efficiently identifying customer requirements.

3. **Smart management:** In the new intelligent branches will be a huge amount of data and IT devices in intelligent branch, putting immense pressure on IT operations and management. The Intelligent Operation Center (IOC) platform implements unified, centralized management of O&M of all branch devices at the bank's central branch. The Internet of Things asset management system that will be deployed should provide full life cycle management, from asset distribution and monitoring to recycling.

4. **Smart/intelligent security protection**: Security protection is critical to the financial industry, and intelligent security protection is essential for branch building. Advanced AI, algorithms, an array of smart cameras, video analytics platforms, and surveillance capabilities such as behavioural analysis, mask identification, and detection of loitering, stalking, theft, disasters, and fires. Also, detection of internal controls such as heat maps within a store, monitoring when employees leave the building, access control and attendance.

A One-Stop Financial Service Centre

With the rapid technological development, digital transformation, smart banks have become inevitable. In light of this, commercial banks, in order to gain a foothold in the market and in the face of intense competition, must pause for a moment and tackle the challenges head on.

When commercial banks digitally transform themselves, they can provide efficient and convenient services to their customers while achieving high-quality development for themselves.

Soon, competition among commercial banks will no longer be limited to business forms, products, and interest rates. Instead, banks will focus on FinTech, 5G, AI, data management, and security.

In the face of increasingly homogeneous competition, traditional banks will need to leverage technologies such as AI and 5G to overcome technical barriers, risky loopholes, and other issues that arise in smart transformation. Only by radically smartifying financial services can they continue to grow and embrace the future.

Every bank's sacred goal on the financial market is to develop bank branches into one-stop financial service centres. Branches no longer provide traditional financial services. New application scenarios and devices AR & VR shopping, car finance, instant payment, intelligent customer service, robots to a remote advisory, capsule-style mobile financial services, and coffee banks are emerging.

Artificial intelligence, combined with 5G and cloud platforms, will provide strong technical support to make highly-automated intelligent bank branches a reality.

AI and VR-based Smart Banking

Within the next ten years, 80% of customers are expected to use AR, VR, or mixed reality applications to see how a product will look before they purchase it, according to Futurum Research. It will take years for this technology adoption, but tech companies are already working on developing it. There's no doubt that AR and VR will take the customer experience to a brand new banking level.

In the past decade, big tech companies like Google, Apple, Facebook, and Amazon have actively shaped customer expectations. An intuitive experience of social media applications or e-commerce sites is expected from any service, including banking.

Customer demands improve every year, as do the technological possibilities to meet these expectations and even over-deliver. The big tech companies are exploring ways to use digital technology to provide customers with a new kind of experience.

To keep up with technological advancements and meet rising expectations, banks are becoming the new pioneers of digital banking reality. AR, VR, virtual assistants, chat services, the Internet of Things, robots, and neuro-services will soon be an integral part of our daily lives.

Some banks and FinTechs are already triumphing over their customers by demonstrating VR/AR features in their products. For example, the FinTech "Acorns" offers a debit card with an AR engagement that can be viewed via smartphones. At the same time, Westpac Banking Corporation offers financial visualization and budgeting via smartphone-based AR. The level at which the technology is implemented is still in its early stages.

There are still significant technical challenges that need to be solved to start a widespread VR/AR device distribution. To create genuinely AR glasses that would be easy to wear every day (compact, light, and easy to use as everyday glasses) it's not so easy.

The technical functionality of virtual and AR glasses is similar to a smartphone. They should contain a battery, screen, a microprocessor, and a modem (5G?) that must be fitted into the glasses. Energy consumption of glasses will be significantly higher because currently available VR glasses consume up to eight times more electricity than computers. Therefore, the implementation of it will require much higher capacity batteries.

We can expect the market to be disturbed with the same impact as it was with smartphones only 15 years ago. That's why the ability to adapt becomes critical if we want to conquer the digital age.

Today, VR sets are used mostly for gaming. This demonstrates that consumers are hungry for this brand new experience.

Virtual or AR will not immediately replace online current banking solutions. That's why virtual and AR products should be easily adaptable for multi-channel use.

How It Will Impact the Digital Banking Experience?

With boosting network capabilities, 5G is opening new horizons for developers of finance-banking software, enabling them to deliver to life projects previously held as well as several new ones (Figure 4.3).

Customer experience is undergoing a significant transformation with 5G-enabled services to customers. Given its improved bandwidth and ultra-low latency, 5G would make it possible to experience AR and VR on smart devices (watches, phones, tablets, smart-rings), taking customer experience to the next level. Companies would be able to deploy in-branch 5G-enabled robots capable of speaking with clients and assisting them with monetary

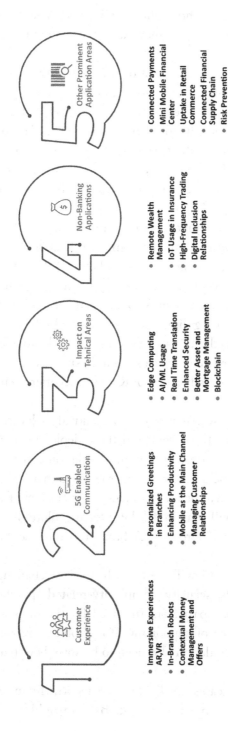

Figure 4.3 Impact of 5G.

transactions. Big data and mobility will make it more straightforward to push marketing or remote advisory content to customers in moments of need.

Improvements in communication and boosting the growing use of smart and wearable devices for financial transactions. Seamless video connectivity will help managers quickly address customer questions and requests. It would also improve firms' productivity by connecting branches, call centres, and agents – using video-based services. Better IoT connectivity would also help firms deploy solutions that enable clients to navigate through offices and receive personalized greetings (KYC).

Payments will be embedded into any device, and the friction in the payment process will be significantly reduced. With reliable connectivity, carrying physical wallets will become a thing of the past. Ecosystems of autonomous payment devices may interact with internet-connected devices to automate account management activities. The emerging digital market would also be boosted as 5G enables personalized shopping experiences and provides customers' offers on their smart devices. Brands such as Amazon, Zara, and Alibaba have already added AR modes to offer their next level experience.

5G is currently available only at a handful of locations and to selected customers. The coverage of these deployments is deficient and mostly aimed at areas with high population densities. 5G networks encounter the final set of tests to ensure that it is ready to be launched. Soon it will be an introduction of 5G New Radio (NR). Many networks will remain 4G long-term evolution and increasingly move towards "pure 5G". A broad roll-out may take at least five years.

The benefits of 5G will come with a set of challenges related to adoption. Costs, security, complexity-related apprehensions on the adoption front, apart from the need for the mass adoption of devices, are capable of supporting 5G. Problems related to standardization and regulation also need to be sorted out before a mass roll-out.

While the use cases are likely to transform financial services, it is not going to happen overnight. Banks would need to take a multi-year approach to build 5G solutions and deploy them out worldwide. It is unlikely to be a universal network before 2025. It is estimated that by 2023, less than 20% of the world's population will have implemented access to 5G. First movers will be Europe, the United States, South Korea, China, Japan, Canada, and Australia.

Therefore, financial institutions would have enough time to build robust strategies and execution plans to roll out 5G-enabled services and remove them from previously mentioned monolithic architecture.

KYC Onboarding

Knowing your client or know your client (KYC) in financial services requires that experts make an effort to confirm the identity and risks involved with having a business relationship. The procedures fit within the broader scope of a bank's Anti-Money Laundering (AML) policy. Banks and financial institutions of all sizes also use KYC processes to ensure their proposed clients, representatives, consultants, or distributors are anti-bribery compliant and are actually who they claim to be.

Banks, insurers, export creditors, and other financial institutions are increasingly demanding that clients provide detailed due diligence information. These regulations were initially imposed only on the financial institutions, but now the non-financial industry, FinTech, and non-profit organizations are liable to oblige.

Banks have to comply with AML and KYC regulations in client onboarding processes. According to AML and Know Your Customer KYC laws, Banks must apply a risk evaluation to their new clients. AML check and KYC check have to be done to comply with AML and regulations.

These controls' objective is to enable Banks to detect possible risks of their clients and execute control mechanisms suitable for clients' risk levels. AML controls include penalties against financing terrorism, money laundering, or conducting business with blacklisted countries. Banks must implement multiple measures in cases such as identifying risky payments and dangerous client activities. If these regulations are not followed, banking institutions will be penalized.

In today's fast-paced world, clients need to get immediate access to products and services. A smooth remote onboarding process helps banks create excellent clients experience while fulfilling regulatory requirements and security. To fulfil those requirements, it is necessary to:

- verify clients identities in a matter of seconds
- reduce manual error risk by using human power
- improve onboarding rates
- effectively prevent fraud
- compliant with AML KYC regulations

Minimizing the friction during the KYC and onboarding process can be done using current technology (3G and 4G + smartphones). ID digital mobile onboarding software creates today a smooth user experience, guiding clients through the entire verification process, which can be done by several steps described on example (Figure 4.4).

In November 2018, the Federal Reserve issued a joint declaration supporting banks to become more sophisticated in their approaches to identifying suspect activity and experimenting with AI and digital identity technologies. Also, the European Supervisory Authorities pushed new solutions to address specific compliance

Step 1	Step 2	Step 3	Done!
User photographs their ID from both sides	User takes self-portrait with their smartphone camera or webcam	User performs some simple movements for liveness detection purposes	User is verified, onboarded, and ready to go

Figure 4.4 Digital onboarding process.

challenges. They suggest maintaining a common approach for consistent software standards across the EU.

Those authorities anticipate several types of control. One of this control is software that automatically identifies and verifies a person from a digital image or a video source or a security feature that can detect images that are or have been tampered with whereby such images appear pixelated or blurred.

How do you know someone is who they say they are? Customer Identification Program (CIP) is designed to limit money laundering, terrorism funding, and other illegal activities. Different jurisdictions have similar provisions; over 190 jurisdictions worldwide have committed to recommendations from the Financial Action Task Force (FATF), a government organization designed to fight money laundering.

These recommendations include identity verification procedures. The wanted outcome is that obliged entities accurately identify their clients.

A crucial element to a successful CIP is a risk assessment at the bank level and the level of methods for each account. While the CIP provides guidance, it's up to the bank institution to define risk and policy for that risk level.

The minimum requirements in the CIP are:

- Name
- Date of birth
- Address
- Personal Identification number
- ID Card/passport document or any other government-issued document

The bank must verify the account holder's identity while gathering provided information during account opening.

Procedures for identity verification include:

- Documents.
- Non-documentary methods (these may consist of comparing the customer's information with consumer reporting agencies, public databases, among other due diligence measures).
- A combination of both.

These policies should be followed. They need to be clarified and systematized to provide continued guidance to banks, financial institutions, and regulators.

The exact policies depend on the risk-based approach of the financial institution and may consider factors such as:

- types of accounts offered by the bank
- methods of opening accounts
- types of identifying information available
- the bank's size
- location,
- customer base, including the types of products and services used by customers in different geographic locations

In a Nutshell

What Will Be Good to Take from This Chapter?

Edge Computing   Edge computing is a model for distributed computing that brings computation and data storage closer to where they are needed to improve response times and save bandwidth.

The roots of edge computing lie in content delivery networks created in the late 1990s to deliver web and video content from edge servers close to users. In the early 2000s, these networks emerged to host applications and components on edge servers, leading to the first commercial edge computing services that hosted applications such as merchant locators, shopping carts, real-time data aggregators, and advertising insertion engines.

Smart Banking   Smart Banking means a more innovative approach to customer service. An example of this is Virtual Greeter: a service that allows customers to access multiple products from a kiosk in the branch or a mobile device.

The Smart Banking initiative enables banks to optimize customer engagement models to drive customer loyalty with high-quality service experiences, attracting new customers with a redesigned sales experience and ultimately engage more with customers.

KYC   KYC indicates Know Your Customer and sometimes Know Your Client.

KYC confirmation is the mandatory process of identifying and verifying the identity of the customer when an account is opened and periodically over time.

In other words, banks need to ensure that their customers are really who they say they are.

Banks can decline to open an account or terminate a business relationship if the customer does not meet the minimum KYC requirements.

•

5

Digital Consumer

A wise person should have money in their head, but not in their heart.

— Jonathan Swift

Till the end of the first decade in the 21st century, the consumption relationship was almost unilateral. The only interaction between brand and clients happened at the time of the sale, exchanging merchandise, or service for money.

The digital age changed this and transformed this reality, and consumers have come to have much more power. They want to be heard, and their opinion needs to be taken into account in companies' digital strategy. Consumer surveys have increased every day, seeking to understand better the digital consumer's characteristics and better meet their needs.

We live in a reality where information is widespread and quickly found. Search engines like Google and others have made prior research the first step before deciding on product or service purchasing. New surveys point out that about 90% of consumers conduct an in-depth study on the product before making the final decision.

And this is very easy, doable in just a few clicks. Now, it is possible to research prices, characteristics of products or services, the opinions of those who have already bought a product or service, complaints, and everything relevant long before you swipe a credit card.

Future of Money

Cash is in the free fall phase. The editor of New Money Review Paul Amery claims, "The death of cash has been pushed further into the future as governments intervene to keep it alive, but they are only delaying the inevitable". He predicts its abolition by early

DOI: 10.1201/9781003198178-5

2030, with central banks in several countries working on centrally bankable digital currencies and fears of losing control of the monetary system.

Bank of England has launched a discussion highlighting the scope for a currency to be used as an unconventional monetary policy tool. China is doing more to grow the future of digital money by launching a digital yuan. China will potentially become a cashless society within the next few years. For most worldwide countries, for national security and equality reasons, governments will allow cash to disappear.

Application Programming Interfaces will shape the future of money, creating a hyper-personalized ecosystem in which consumers' needs will be predicted.

Edinburgh University's Speed says the United Kingdom's open-banking revolution has not been in vain. People now have third party relationships and use an open banking protocol. The Financial Conduct Authority is working to make open banking more like a safe garden. Databricks' Nakai says: "Open APIs create opportunities for banks to make much more holistic decisions about individuals, making banking much more personalized and inclusive".

Amery of New Money Review adds, *"We should stop thinking about banks and technology companies. Money is now inextricably linked to technology. But money will also become increasingly confusing and annoying".*

Technological advances mean financial and banking services are no longer the domain of banks. Banks are at risk of becoming services with low brand awareness, little differentiation, and diminished customer loyalty (Figure 5.1).

Banks' future looks bleak; most of today's banks are closing branches and sacking staff as they struggle to stay above the surface with low-interest rates and the switch to digital. Just by comparing HSBC Bank shares and FinTech Revolut shares over the past year, HSBC Bank has seen its share price drop by 25%, and today Revolut's valuation has more than $5.5 billion.

Non-cash payments continue to increase. By 2020, the total number of cashless transactions increased by 8.1% to 98 billion in the eurozone area. Half of these transactions were made by credit card, followed by credit transfers and direct debits.

Figure 5.1 Revolut.

As we entered the digital age, the nature of goods and services is changing quickly. The same is with the money. Digitalization and technological advances are accelerating the process of dematerialization. People want an instant transaction, services-on-demand, and goods available at the moment of purchasing.

The COVID-19 pandemic has accelerated this digitalization trend, with a surge in online payments and a shift towards contactless payments in shops.

New forms of private money have emerged to meet digital payment demand. They are available as retail bank deposits, which can be used for transfers, including direct debits and electronic money through cards and mobile payment apps.

The Eurosystem administration mechanisms ensure that commercial banks and payment service providers are effective and safe in the eurozone. This empowers people to continue to trust private money, which remains an essential part of our financial system.

But money in digital form from Central Bank is still not prepared for commercial payments.

The Digital Euro

The European Central Bank wants to guarantee that the Euro stay a good fit for the digital era. In early 2019, the European Council decided to explore the possibility of issuing a Digital Euro.

The Euro Council evaluates the implications of the potential introduction of a Digital Euro. In October 2020, the European Central Bank published the report on a Digital Euro and started a public discussion.

Government money is unique. It gives people unrestricted access to a risk-free and trusted means of payment that can be used for any necessary transaction. But for retail use, it is only possible physically in the form of cash.

A Digital Euro will complement cash and guarantee that consumers continue to have unrestricted access to government money in a form that meets their digital needs. It could be significant in various future scenarios, from a drop in cash usage to pre-empting the uptake of digital foreign currencies in the Euro area. Creating a Digital Euro might become necessary to ensure continued access to government money and monetary freedom.

A well-designed Digital Euro would create synergies with the financial payments industry and allow the private sector to create new businesses based on Digital Euro services. A Digital Euro would also unify Europe's digital economies of all countries in the Euro zone.

Cryptocurrencies Show Risks

Innovations like distributed ledger technology, particularly Blockchain, bring new opportunities and a lot of unknown risks. Transactions between peers happen directly, with no need for a trusted third-party intermediary. Cryptographic proofs replace the trust that is usually inherent in a transaction. The security and integrity of records are ensured by distributed ledger technology, which avoids the double-spending problem.

The main risk lies in relying completely on technology and the flawed concept of no identifiable issuer or claim. This also means that payers cannot rely on crypto assets to keep a stable value: they are highly active, speculative, and do not fulfil all the functions of money.

If widely approved, cryptocurrencies could endanger financial stability and monetary sovereignty. For example, if the issuer cannot guarantee a fixed value or is recognized as incompetent of swallowing losses, a run could occur.

As the economy evolves and new expectations about the money emerge, the banks must be ready to respond and ensure that world payments' financial institutions adapt to changing consumer preferences and remain inclusive and efficient.

Despite all the changes we mentioned, the grounds of money remain intact. People accept money only if it is extremely trusted, has its value, and respects privacy. These rules have been and will continue to be found in government money, regardless of its form in the future (Figure 5.2).

Cutting Banks as a Middleman

From PayPal to mobile payments to other electronic payment industry segments, convenience is battling with transmittal fees. On the one hand, clients like the idea of sending payments digitally rather than paying in cash or using a credit card. However, global financial

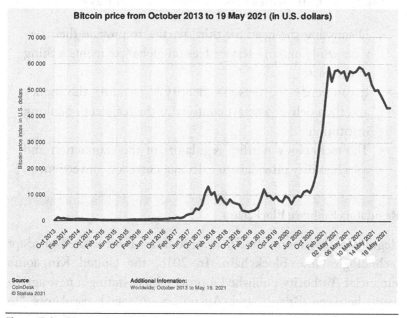

Figure 5.2 Bitcoin prices oscillation.

institutions have consistently been waylaid by banks and third parties that tie significant fees on top of the payment amount.

These fees are billed in most cases as percentages. How high can be those percentages? According to the World Bank, the average payment fee is 7.5% for consumers and 10% for commercial entities. Those percentages are more extensive for smaller payments. Get into more significant transactions and these transmittal fees can become very expensive. The appearance of third parties in the global payment market prevents the financial industry – and consumers worldwide – from embracing the ease of digital transactions.

Blockchain cuts the banks and financial institutions as a middleman out of digital transactions, among other benefits. Digital ledger technology kills a lot of the labour and processing involved with payment transfers. Third parties' fees to verify the transaction's validity and security make conversions from one currency to the next and complete other essential processes. There is a reason that transmittal charges are shown as service fees: third parties need to provide services to make the payments go through, and they are paid for those services based on the amount of the payment.

Blockchain simplifies this process in several ways:

- Dismissing the need for third parties to provide their services while making service fees on global payments a thing of the past.
- It can verify the person's identity by sending the money and then deliver a smart contract to the recipient's banking institution.
- The technology notifies regulators of the transaction and uses liquidity providers to calculate currency conversion.

Regulators Embrace Blockchain

Regulators are among the financial industry most excited to adopt technologies like Blockchain. In 2016, the United Kingdom's Financial Authority published a press release stating a new agreement between British and Australian financial regulators "to encourage innovative businesses. In essence, the deal made it easier for finance innovations to take root across oceans and boundaries, which could be of enormous benefit to the problems attached to global payments.

Blockchain by name was mentioned when discussing FinTech business models "asking assistance about how to operate the regulator requirements". Similar agreements exist between the United Kingdom and Singapore and Singapore and Switzerland.

The Future of Digital Transactions

DLT takes care of everything from the moment the sender submits a payment to the moment the money arrives in the recipient's account on the other. Blockchain technology can make digital payments faster so that a payment recipient can access funds just seconds after they are sent (Instant Payment). No longer will it take days to process money transfers. And since Blockchain makes payments more secure – with protections against fraud, scams, and more – it is likely it will only be a matter of time before it becomes standard in all worldwide digital transaction payments.

Swift recently launched a "global payments innovation" service or GPI. Opened in May 2019, the GPI service connects 12 banks with its presence in 60 countries. Without DLT digital ledger technology, this network allows live international transactional payments as well as payment tracking.

Payments also aren't instantaneous, as they would be with DLT, but Swift explains they will only take a day to process through GPI instead of the old three-to-five-day standard.

In a Nutshell

What Will Be Good to Take from This Chapter?

Cash is in free fall. The editor of New Money Review Paul Amery claims, "The death of cash has been pushed further into the future as governments intervene to keep it alive, but they are only delaying the inevitable". He predicts its abolition in the early 2030s as central banks in several countries work on central bankable digital currencies and fear losing control of the monetary system.

The Bank of England has launched a discussion highlighting the possibilities of a currency as an unconventional monetary policy tool. China is pushing the future of digital money with the introduction of a digital yuan. China could become a cashless society within the next few years. In most countries around the world, governments will make cash disappear for reasons of national security and equality.

The Programming Interfaces application will shape the future of money, creating a hyper-personalized ecosystem where consumer needs are predicted.

6

SYNERGY

The whole is greater than some of its parts

– Aristotle

The association of technological synergy and the performance of new products or even their characteristics is a widespread phenomenon that is universally accepted with little to no resistance or objection. But in terms of concrete evidence to support this notion – for example, research papers – there are little to none that integrates these associations. But so far as the traditional approach and understanding go, we can't deny that the technological synergy has a positive effect on the performance of not just products but entire industries as well as a strong and positive effect on the advantages of those products to be explored by their relative industries.

Additionally, there has been observed an indirect effect on the performance of the new products by the virtue of product innovation as well as advantages associated with those products. In essence, we can say that product innovation, as well as the exploitable advantages, act as powerful intermediaries between the performance of the products and the technological synergy they create.

The question at hand then becomes that is there an ideal way for the use as well as the organization of the technology? Is it to be limited to a particular sector/s or are their applications further explorable as we have seen in numerous cases in the past. In the context of the development of online banks and FinTechs, can we guess whether there is more competition or synergy? It is possible that in the likely future, the development of online banks and technology companies that are working to develop their own financial

DOI: 10.1201/9781003198178-6

services similar to FinTechs may be executed in parallel to each other. And as evident, it will likely lead to the development of a synthetic model comprising of numerous methodologies to combine the core concept of FintTechs and banks. This will make it blur to identify who has actually taken over whom. Whether the FinTechs or the electronic banks have emerged victorious remains to be an opinion based on personal believes and the evidence which may be biased in one direction more than the other.

Digital Footprint

In many ways, technology has ceased simply to be a tool to assist organizations or provide better customer service or even speed up production in bulk quantities. Progressively, the organizational structure of banks and many other organizations are being revolutionized by technological innovation. But looking at the big picture from another perspective, the executives and managers are faced with problems of their own (Figure 6.1).

How It Works

As we have previously stated, digital footprint basically works by collecting information on all your online activities and communications. These can be both, active as well as passive activities.

Examples

Examples of active activities that leave behind a digital footprint include using social media platforms and posting images, status, or even commenting. Passive activities include other websites collecting your IP address and even accepting cookies. Whether you are placing an order online, logging in, registering, or conducting an activity online that requires you to present any form of identification or to accept something from the website, or leave a mark, you have left behind a digital footprint.

Test by googling your name (Figure 6.2).

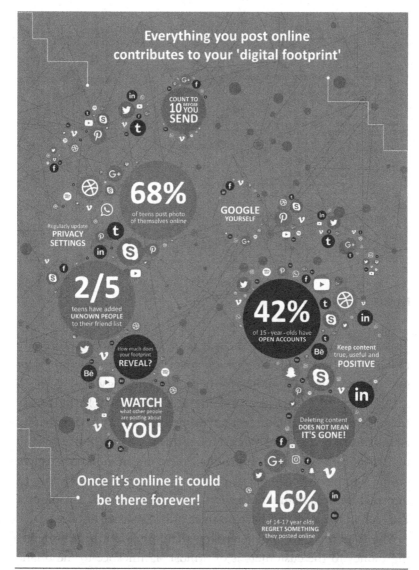

Figure 6.1 Digital footprints.

Understanding How It Functions

The implementation of rapidly evolving technology at the pace it is evolving is nothing short of a challenge. The selection, management, and implementation of the technology changing rapidly are giving birth to the requirement of technologically competent managers.

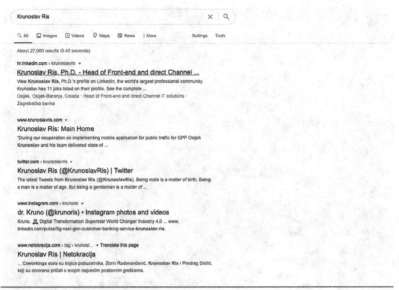

Figure 6.2   Search yourself.

This again calls for shifts in the way management processes are carried out, for they call for higher efficiency as well as the appropriate selection and implementation of the right technology.

As the internet grows, we can find traces of easy, simple, and accessible information about almost anything we want. Whether that be an individual or an entire organization, this is what we refer to as the digital footprint. If for an individual, mere registration on a simple online forum or a social media platform can leave behind chunks of information, it is highly unlikely that an entity such as a bank or a FinTech in the least will have no trace of its existence online.

The importance of understanding the digital footprint for lending money to the customers is of utmost significance to the financial organizations whether they be FinTechs or digital banks. On another note, many regulatory bodies don't even recognize electronic banks as they are; rather, they are just banks conducting their activities online despite there being numerous prominent technical differences.

The digital footprint is a powerful tool for such organizations to develop a track record of their client base. Since it helps in collecting even minute details such as whether the client is conducting the financial transactional activities on iOS, Android, or their computer, the channel that the customer uses to access the website (search engine, direct access, referral, or any other medium), etc.

Why It Matters

As we see remarkable changes being made in the way payments are made and transactions are processed, the time demands that the system that assesses this information is also upgraded. In this era of digital banking, there is an ever-increasing risk of digital fraud. When traditional credit scoring fails to serve the intended purpose, a digital footprint can help and make up for missing data.

We can include cookies, online patterns, and even information from social media to track personal financial behaviour. Factors such as browser used, purchase history, and IP address/es are all useful to compose the digital footprint of an individual and help detect and even prevent fraudulent activities.

One of the most useful applications of digital footprint would be the use of information from the internet usage pattern of an individual for the collection of information to establish who they are. No denying that adoption of digital mediums is only going to keep increasing, it is now considered an invaluable source of useful data especially for regions with a high number of citizens involved in banking activities.

Some may argue that the use of alternative data to make up for digital footprint may seem like going an extra mile, but it is helping the FinTech industry in numerous ways to reshape itself for the better. We now have new and better credit score evaluating algorithms that can easily multivariate and also structure, mine, calculate, and weight data to further enrich it and use it for futuristic credit scoring. Many organizations, whether they be of the banking sector or simple FinTechs, are using similar and even further complicated data sets to assess the creditworthiness of the consumers who are indulging in their services; especially the ones applying for loans.

Data Centralization

Nowadays, there is too much data flowing in all directions and being generated from all directions. It is now so abundant that the possession of its excessive amount is not even a factor. What matters now is the way it is utilized. This data is proving out to be a game changer in a number of industries ranging from applications such as enhanced data security, improved business processes, and improved performance of online banks and FinTechs.

The age of technology requires that financial institutions manage their information by learning from their experiences and feeding the extrapolated data into their agenda for strategic development. It is only with the proper management, which includes:

- The collection of data
- Its recording
- Its efficient monitoring
- Cleaning
- Visualization
- Analysis
- Interpretation

The data can be fully explored to its complete potential. This requires the formulation of a proper data strategy and its reinforcement so that the value of data can be established as a fundamental asset to the banking and finance business.

In the modern context, data is being referred to as a new commodity. It possesses characteristics (like scarcity) that are similar to traditional commodities. Do keep in mind that data is not like information, which exists in abundance. For information to become data, it needs to be collected, recorded, and safely stored first for it to become data. It is after this that it can be benefited from. Additionally, the entire process for information to transform into data is actually quite cumbersome. This is why data is considered a scarce resource, for even if the information contains data, the process of extrapolating that data from that information is what makes data truly valuable. Not to mention its applications afterward.

Since it is such a precious asset, financial organizations see it as a highly valuable resource for their sustainability and profitability. Banks already possess a tremendous amount of data about their clients in the form of their transaction records, this includes transactional history, bank ledgers, and channel traffic. Besides the internally existing resources, there are various external sources for data available as well which are available to benefit both, banks and FinTech organizations.

Coming to the presence of big data and the centralization of data in the finance industry, though both the industries have the objective of saving the data of their clients at the core of their operations, there are numerous technological and legislative challenges happening globally that have made it even bigger a challenge. The information that is possessed by or available to FinTechs and banks has had its security tightened to such an extent that it is now being censored from marketers who had previously not only access to it but benefited from it as well. The centralized and collectively stored set of data is protected to such an extent that marketing tools that make personalization and data segmentation possible in other industries in often off limit to institutes operating in the said industries.

Defining Data Centralization

In simple words, data centralization is the process of collecting, storing, and maintaining data in a single location. But at the same time, the data is made available to be accessed from more than one location, ideally, anywhere in the world (Figure 6.3).

Benefits

This results in numerous benefits for the financial organizations looking to have their data stored in a single and remotely accessible location such as:

- It allows for quick searching through the data. As all of it is stored in a single location, the search engine needn't search through multiple sources to collect the desired results.

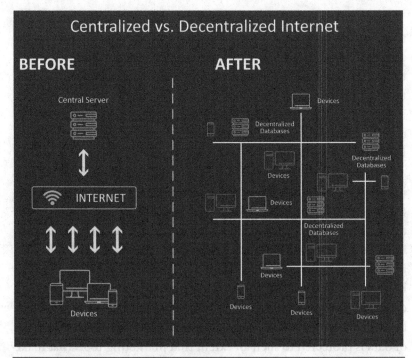

Figure 6.3 Data centralization/decentralization.

- It accounts for a simplified data structure which eliminates the pitfalls that are commonly associated with complex IT infrastructures such as high cost for installation, added cost for security demands, difficulty to understand the complex operations of the infrastructure for average workers, and an increased cost of maintenance and repair. But with centralized data, the overall system is streamlined and all the above-mentioned issues never take place.
- Data centralization allows for accurate and up-to-date information. If the data is stored in multiple locations, changes in data in one location will require changes to be made in the data present in other locations separately. This is not the case with a centralized data system.
- This also means that redundant data can be eliminated centrally which eliminates the possibility of confusion sprouting. And of course, all of the above-stated benefits allow for efficient collaboration between the teams despite their locations being far apart from each other.

For financial institutions, data signifies the track they must proceed on and without it, they may be lost on which path to take next. Fragmented, incompatible, and incomplete data stifle innovation and make organizations less innovative, efficient, and competitive. So centralized data helps organizations optimize their data assets, enhance consistency, help eliminate distractions and allow them to focus on what's more important, and enhance the customer experience. As a result, the organizations can not only save money but time as well.

Numerous case studies have indicated that data centralization leads to meaningful reporting, increasing the organizational ROI, and ensuring that similar data is available across the organization.

Cloud is one of the most thrown around terms when we talk about data centralization. Cloud computing has actually revolutionized data centralization and allowed for the establishment of data centres that now host millions of terabytes of data for different organizations.

Quantum Computing

Coming from data centralization to the phenomenon that equally perplexed and intrigued the world, quantum computing is being viewed as a gateway for new possibilities and crossing the threshold that was previously limited due to the traditional computing capabilities. So, let's start with how quantum computers work in the first place and what makes them so different from traditional computers.

How Quantum Computers Work?

Quantum computers perform calculations that are based on the probability of an object's state before it is measured in 0 or 1. This means that they have the potential to process exponentially more data in comparison to traditional computes.

Logical operations that are carried out by the traditional computers are based on logical operations and a defined state which is the binary digits. This can also be referred to as a single state which is represented as 0 and 1, also called on and off or up and down. But quantum computers use qubits which are generated from the quantum state of an object. These quantum states are the unde-tected and undefined properties of an object such as the polarization of a photon or the spin of an electron.

Instead of having a defined position, the objects that occur in the quantum state exist in a mixed position called "superposi-tion", something akin to a spinning coin. Although the superpo-sition of qubits is an amazing property, the entanglement is what makes them of use to us as it is easier for entangled qubits to interact instantly. But to make functional qubits, quantum computers need to be cooled down to absolute zero. But even then, the qubits can't maintain their state of entanglement for long.

Quantum Computing and Banking

The finance industry is known to have repeatedly made use of tech-nology and even complex physics and mathematical concepts to ensure that it is reaching faster and more secure processing. The financial transaction makes use of complex algorithms which are always in the need of being made faster and efficient. This is seem-ingly possible with quantum computing. As a result, the institutions will be able to scale their processes and reduce their costs by help-ing them to replace their Human and IT resources. This is a viable prospect for organizations that are on the lookout for boosting their data speed, trade, and transactions much more securely.

With the flexibility of qubits, it becomes possible to study a qubit in its superposition state. This also allows for saving more data in comparison to the traditional bit. The founder of Guggenheim Partners' Quantitative Investment Strategies and CEO of True Positive Dr. Marcos López de Prado has stated that the importance of quantum computing in the world of banking and finance will become a necessity over time. He also stated that probably in 20 years or so, quantum computing may not only be just important, in fact, it may be the only ideal solution for us from both computa-tional and energy perspectives.

Though we may still be a decade or so away from the earliest com-plete adoption of quantum computing in the banking and finance industry, the potential of this technology is being viewed as increasingly lucrative by capital marketers. The possibility of minimizing risks and maximizing profits speaks for itself. The prospect will have several pros and cons for FinTechs all the same just as it will be for banks and similar institutions. Critical areas such as cybersecurity and data encryption can also benefit from the inclusion of quantum computers. It can help in the early and timely detection of fraudulent activities with the identification of behavioural patterns that are not possible for normal computers.

Cyberattacks and threats from hackers are a growing concern as the internet becomes more and more connected. But to have financial data encrypted with quantum cryptography, the level of security can be raised significantly such that it will not be possible for hackers to get past the security that is fortified by quantum computers using only traditional means as the data that is encoded in the quantum state cannot be read or even accessed without using quantum means.

In a statement, Dr. Bob Sutor who is the vice president of IBM Q Strategy and Ecosystem stated that a lot of advanced and biggest organizations in the world are exploring the prospect of how the early development of classical-quantum algorithms will provide a competitive advantage to their organizations against the competition they face in the market. All in all, the impact of quantum computing is being predicted to be a predominantly good thing. The implications will be far-reaching, to say the least.

In recent news, a statement given by Deltec Bank in the Bahamas stated that there are banks that are actually trialing quantum computers and have successfully solved such problems that were previously highly resourced intensive or impossible to complete. Besides the financial sector, some other industries such as telecom, manufacturing, security, and many others are predicted to experience a high level of disruption.

An article from October 2019 that was published in Banco Bilbao Vizcaya Argentaria (BBVA) suggested that by the introduction of quantum computing in the banking sector, perhaps the banking we are familiar with may be changed forever.

An Example

We know that for the entire banking and Fintech sector, it is vital to have the personal data of the clients encrypted so it is not easily readable for a hacker. The highest standard of encryption that is employed by the banks today is **RSA-2048**. If we were to use a powerful classical computer, it would take it about 1,034 steps to decrypt that algorithm. Think of it like this; if a processor was capable of carrying out 1 trillion operations per second, it would still take it 317 billion years (more or less) to decrypt the information. Of course, that just makes it impossible to decrypt the information in realistic terms.

But if we were using a quantum computer for this purpose, with the power of superposition and entanglement, the qubits would allow for the same encryption to be decrypted in merely 107 steps. If our quantum-powered computer was running at just 1 million operations per second, it will only take about ten seconds to complete. But these are just estimates and figures. We are still a little way away from reaching that point but we are getting there a bit by bit.

We even have the likes of Barclays and JP Morgan that are working with quantum computers to speed up their financial transaction and making them more secure. What would take traditional computers years to calculate is predicted to be calculable by quantum computers in the brink of just a few seconds.

5G Fundamentals for 6G

It seems as though we were just getting acquainted with 5G that we are suddenly bombarded with the new 6G coming in the market. Though some countries remain to be upgraded to 5G, the prospects that are offered by 6G greatly surpass those that were offered by its predecessor.

The Basics

We will make it simple enough to be understood by the majority. We know that modern communication is carried out via electromagnetic waves that are characterized by two main features:

- Frequency
- Wavelength

The communication devices we use have transmitters and receivers that are used to transmit and receive those electromagnetic waves. We will start with 4G that is almost globally present. 4G wireless network runs on millimeter waves and on the spectrum that is based between low and midrange i.e., a bit under and over 1 gigahertz. This means 1 billion cycles per second. This was beaten by 5G that took it up to 300 gigahertz or 300 billion cycles per second.

With such high speeds, it became possible to send information-dense data (videos, etc.) relatively quickly. In essence, the jump from 4G to 5G represented the expansion of the spectrum that was used for communication and the introduction of new frequencies. We basically moved from bigger and more visible wavelengths to smaller and less visible ones.

But 6G beats its predecessor by three times; it transmits data three times faster at 1 terahertz or 1 trillion cycles per second. In other words, this basically yields a data rate of about 11 gigabytes per second with experts estimating that it could even go up to 8,000 gigabytes per second. In comparison to 5G, they will have much higher bandwidth and much lower latency.

The Challenges

Just as there were numerous challenges with 5G, there will also be many with 6G. One such challenge is said to be the development of commercial transceivers that will operate on the high terahertz frequency of 6G. This puts a lot of stress on the electronic component providers as they will have to work on several key areas. For example, the manufacturers for the semiconductors will need to deal with extremely small wavelengths, and corresponding to which, they will have to deal with the physically small size of RF transistors and how they will be working in correspondence to the element spacing of the terahertz antennas arrays.

Additionally, the need for radio access points for 6G will be significantly larger than that for 5G. In other words, the impact on the Radio Access Network (RAN) will be substantial as the change is largely driven by an increase in frequency which requires an increase in the numbers of antennas. As the terahertz waves fall between the micro and infrared waves, it is difficult and expensive to generate and transmit them. It creates a need for a special laser and even after that, the frequency range remains limited.

Developments and Solutions

The team from Nanyang Technical University (NTU) Singapore that developed the chip which can transmit terahertz waves at ultra-high speeds which are going to become the basis for 6G used a new material to make the ends meet. The material is called Photonic Topological Insulators (PTIs) and it can conduct light waves on its surface and edges which eliminates the need to have the light run through the materials and allows for it to be redirected across corners without disturbing its flow.

In a sense, we can say that 5G paved way for the majority of this convergence to take place. This includes:

1. The need for deploying edge computing to ensure overall throughput and low latency for URLLC solutions; and
2. Supporting mMTC in the form of many low-power IoT devices that do not have their own onboard computing.

The addition of Mobile Edge Computing (MEC) is considered an addition from the 5G network that will likely be as it is built into the 6G. Core and edge computing are going to be like become much more seamlessly integrated as a collective part of the combined computational infrastructure framework.

Experts have also speculated that the evolutionary path driven by 5G will pave way for 6G and it is quite likely that 6G will continue on along the same trajectory. This includes the use of:

- Highly autonomous networks
- Experience optimization
- Highly improved quality of service
- Service-based architecture approach
- Edge computing

But with technology convergence, we may still get to see some profound core network changes for 6G.

There is no doubt that an integrated set of previously disparate technologies will soon be represented by 6G. So we can expect the convergence of several technologies including:

- Big data analytics
- AI
- Highly complex computation

This and many other potential advantages are being associated with 6G technology as it is soon likely to become active in the coming future. Computational infrastructures will most likely be able to work in conjunction to AI with improved access to AI capabilities. This will also make it easier to determine the best possible location for computation to take place. We are looking here at numerous potential advantages that will likely blossom in all domains of life and across all industries; including banking as well as FinTechs.

In a Nutshell

What Will Be Good to Take from This Chapter?

Digital Footprint   It is a trail of data that you create when you use the internet. This includes the websites you visit, emails you send, and information you submit to online services.

A passive digital footprint is a trail of data you unintentionally leave online. When you visit a website, the webserver records your IP address, identifying your ISP and approximate location.

Although your IP address may change, it is also considered part of your digital footprint. A more personal aspect is a search history, which the search engines store while a consumer is logged in. An "active digital footprint" includes the data that the client submits. Sending an email contributes to your active digital footprint because you expect the data to be seen and/or stored by another person.

The more emails client send, the more digital footprint grows. Since most people store their emails online, the messages you send can easily stay online for several years or more.

Data Centralization   Centralized data means that all your data is in one place, compared to decentralized data scattered across multiple machines and servers. Decentralized data has a higher number of portals. Therefore, security is lower as intruders can enter through numerous access points and obtain corporate data, not to mention the higher cost.

Quantum Computing   Quantum computers are machines that use the fields of quantum physics to store data and execute calculations. This can be highly beneficial for specific tasks, where they could far surpass even our best supercomputers.

Traditional computers, including smartphones and laptops, encode data in binary bits (one or zero). In a quantum supercomputer, the basic unit of memory is a quantum bit.

For example, eight bits are enough for a traditional computer to represent any number between 0 and 255. But eight qubits are enough for a quantum computer to express every number between 0 and 255 simultaneously. A few hundred entangled qubits would be enough to represent more numbers than there are atoms in the universe.

7

MACHINE LEARNING, PaaS, AND OTHER FUTURE BANKING TRENDS

Don't tell me where your priorities are. Show me where you spend your money and I'll tell you what they are.

— James W. Frick

Financial Institutions seeking to boost sales should consider building AI into their strategies. And it should be the top priority. Many examples in the banking sector and beyond show the enormous benefits and endless opportunities AI-driven solutions offer. Think of new revenue sources and increased profitability of existing products, to name a few.

But there is still plenty of room for growth in applying AI-enhanced advanced analytics in financial services.

Another important thing is not to start with the most challenging use cases but choose low-complexity, easy-to-implement applications that offer high benefits. Some banks rely on their tried-and-tested in-house AI tools but more and more swear by teaming up with third-party providers. These companies can help banks marry non-traditional data with traditional information to create deeper customer insights and provide access to AI-driven sales tools that predict customer needs based on behavioural patterns (Figure 7.1).

Financial institutions already use all kinds of AI tools, ranging from chatbots that improve customer experience to mobile banking facial recognition solutions that ease digital onboarding and natural language processing used by AI-powered virtual assistants. M.L. (often used interchangeably with AI, although it is a subcategory of AI) is key in boosting sales through analysing data.

DOI: 10.1201/9781003198178-7

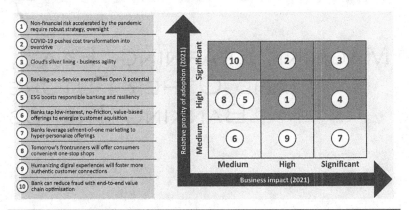

Figure 7.1   Top retail banking trends.

AI Banking Readiness

According to Banking Technology Vision, Accenture, 2017, about 71% of bankers believe AI can become the face of their organization or brand, and 40% plan to invest in embedded AI solutions in the next three years. But the current hype around the use of artificial intelligence in banking needs to be "taken with a grain of salt". Why? The technologies that support AI solutions are probably not quite ready for real-time use as the excitement suggests (Table 7.1).

Table 7.1   Readiness of AI for Banking

AI TECHNOLOGY	READINESS FOR BUSINESS USE	POTENTIAL FOR BANKING
Machine Learning	Medium	High
Deep learning	Low	High
Natural language processing	Medium	High
Natural language generation	Medium	Medium
Visual perception	Medium	High
Virtual assistants	Medium	High
Robotic process automatization	High	High
Robots	Low	Medium

Source: Infosys Finacle.

Learning from Other Industries

Companies outside the financial services sector use AI solutions to influence sales in various ways, from supporting new products to generating leads. About 74% of organizations implementing AI increase sales of new products and services by more than 10%. And more than two-thirds of them also use it to up the number of inbound customer leads, according to a survey by Capgemini.

For example, casino and hotel operator Caesars Entertainment teamed up with a data platform running on Intel systems to process customer data and gain real-time insights for personalization and recommendation engines.

The platform relies on M.L. to help Caesars understand customer journeys based on interactions through reservations, social media, credit cards, and kiosks. It also incorporates third-party data such as construction alerts, weather information, and data from GPS and video systems.

Caesars has shifted to non-gaming expenditures, including entertainment, retail, and dining, but with its AI-driven solution, now it can better understand how and where guests are spending money. The company can directly send each guest promotions and offers based on their interests. The result of the AI upgrade is that Caesars achieved top-line growth and drilled into new revenue streams.

Another example is Cosabella. Cosabella is a luxury lingerie retailer. They moved to a smart managed marketing platform powered with AI platform automates digital advertising and marketing, such as targeting a high-value audience and driving paid search ROI.

Over a three-month pilot, the platform produced a 336% return on ad spend (ROAS) and a 155% increase in revenue. Before they shifted to the AI platform, social media accounted for 5–10% of Cosabella's paid ad revenue. Since the adoption of the platform, it consistently accounts for 30%.

Tech companies, such as Apple, Google, Alibaba, Baidu, and Amazon in the United States, are the biggest consumer of AI. They use it to improve sales through product recommendations, targeted advertising, and demand forecasting.

AI is becoming more significant for tech giants as they are expanding into financial services, posing another competitive threat to banks in most of their key businesses.

Spanish lender BBVA has developed a service recommendation engine to offer clients the best commercial offer based on their most used transactions and navigation patterns.

These data are aggregated in a sorting algorithm, which then creates a recommendation. We need to point that the volume of information is incredibly vast, and the only way to offer a recommendation is using M.L. technologies".

Danske Bank in Denmark is also using M.L. for predictive models to assess customer behaviour and personal preferences and predict the client's needs. Based on online behaviour, Danske identified clients in a specific situation where financial advice is needed. For example, when a client changes jobs with a new salary and pension plan, the bank uses these situations to contact the client and in 62% achieved better results than traditional waiting where the client arrives at one of the branch offices.

Danske Bank's in-house AI start-up, advanced analytics, has also developed a peer-based concept for banking customers, similar to Tripadvisor's model.

It allows the bank to show clients the choice of other people with a similar profile, for example, when choosing pension plans. The concept is named "Others Like You".

Power of Machine Learning in Analytics

Financial institutions, especially banks, sit on top of a hill of clients' data such as social background, purchase history (products purchased and amounts paid), and household income. Thanks to the model shift in sales from being reactive to proactive, these data can be capitalized on, and from instinct-driven to insight- and data-driven.

The algorithm uses data and feedback from people to learn the relevance of provided inputs to a presented output (for example, how the inputs "air pollution" and "interest rates" predict housing prices). There are many different algorithms for the many different business

Table 7.2 Example of M.L. Algorithms

ALGORITHMS	BUSINESS USE CASES
Linear regression	Understand product-sales drivers such as competition prices, distribution, advertisement, etc
Logistic regression	Classify customers based on how likely they are to repay a loan
Linear/quadratic discriminant analysis	Predict client chum and a sales lead's likelihood of closing
Decision tree	Understand product attributes that make a product most likely to be purchased
Naive Bayes	Analyse sentiment to assess product perception in the market
Neural network	Predict whether registered users will be willing or not to pay a particular price for a product

Source: McKinsey.

cases. And algorithms are used in different types of analytics that can be classified based on their complexity or fields of application (Table 7.2).

The advanced predictive and prescriptive analytics can be seen in numerous areas. Banks can better anticipate and predict possible clients' churn and improve the effectiveness of cross-selling activities. The sales industry has only recently begun to use predictive analytics, and prescriptive analytics is still hardly used in the financial and the banking industry (Table 7.3).

Table 7.3 Types of Analytics Options

	FUNCTION	METHODS	EXAMPLES OF APPLICATION
Descriptive analytics	Insight into the past – what has happened?	Data aggregation and data mining	Year over year change in sales average spending by customers, total stock of inventory
Predictive analytics	Understanding the future – what could happen	Statistical models and forecasts techniques, focus on M.L.	Forecasting customer behaviour, purchasing patterns, sales trends, producing a credit score. Detecting e-mail sentiment, lead scoring
Prescriptive analytics	Advise on outcome – what should we do	Optimization and simulation algorithms, focus on M.L.	Optimizing production, scheduling, and inventory in the supply chain, optimizing customer experience

Source: McKinsey.

An excellent example of using AI in analytics is when a bank has enabled analytics solutions to end the migration of high-value mortgage clients to competitors. The company compared the attributes of loyal clients with those that had churned and identified over a hundred factors correlated to clients, products, and data.

The company worked with statistical methods to point to these crucial factors and discovered that city-based middle-aged specialists with gold credit cards tended to churn more than others. The gained data was inserted into a predictive tool that uses neural networking procedures to predict churn behaviour.

The number of factors is reduced to ten, and the model was applied to all mortgage clients, ranking them in order of their possibility to leave. Based on this ranking, the bank started a targeted marketing campaign, which almost half cut the churn percentage.

Machine Learning Options and Sales Strategy

But how to integrate Machine Learning into banks' sales strategy? First of all, collecting better data for M.L. analysis is a critical factor. Using gathered data from a customer relationship management (CRM) system does not always bring the wanted result as these data might be inadequate or flawed. InsideSales suggests that businesses operating with sales personnel should track phone calls and emails to get better data for M.L. processing. This can be the truth for small business operations, but for the banks with a couple of million clients, this is mission impossible.

When it comes to digital sales in banking, without any sales employees involved, thoroughly tracking customer activity in mobile and online banking is essential for measuring client interactions in offline channels. This should be complemented by monitoring marketing contacts and client feedback. All these interactions need to be stored in a central database.

With more and more self-service options (account opening, credit application, mCash, instant payments, or feedback to transaction fraud suspects) popping up by the dozen, banks must put more effort into digital behaviour and digital sales monitor.

Cleansing and digital processing data help derive valuable insights into customer behaviour and purchase patterns.

After tracking clients' activities, M.L. software works with the four categories of resulting data:

- **descriptive** – who the prospects are and what characteristics they have
- **activity** – taken actions
- **contextual** – prevailing conditions, such as weather or economics
- **result data** – what are the outcomes

Machine Learning tools work as exchange engines and can point out which collected data combinations are likely to bring the beneficial outcome. These outcomes can suggest the best channel, frequency, and time to contact the client and what tone should be used during the interaction. All these can help banks create tailor-made offers and a rich digital customer experience. M.L. can also assess marketing interactions' performance, measure effectiveness, analyse customer reaction to marketing messages, and prioritize them based on performance.

Analytics can be used for data-driven and automated sales and guide sales employees by providing M.L. analysis. For instance, a medicines company uses analytical tools to give the field's sales force insight into the overall business, enabling them to create their strategies, implementation plans, and own projects on the platform.

These, of course, could be monitored and tracked by managers. According to McKinsey, within just a couple of weeks of introducing M.L., churn was down and pricing was up. Within a year, it brought in $50 million in EBITDA.

Digital analytics improve the opportunities of reconstructing sales, but hasty investments can be expensive and counterproductive. We also need to mention that sometimes these investments do not seem to pay off at all. Why? Possible culprits include the field salespersons who do not trust the data, overly detailed insights, or sales teams who feel that their own experience and expertise are being ignored.

Machine Learning: The Rise of the (chat)Bots

Some of the banking scene's digital leaders have already started the ball rolling and with considerable success. Here come some financial industry examples, gathered by MarketForce, Bank Administration Institute (BAI), Capgemini, and the non-profit organization advising financial institutions:

- **ERICA – Bank of America** made Erica virtual assistant. Erica uses cognitive messaging and predictive analytics to give 24/7 financial guidance to clients. Erica can make balance check, transactions (transferring money between accounts), view transactions history, and schedule payments. Clients can also chat with Erica through voice or text message.
- **BBVA Valora** – Bconomy is a commercial well-being tool. Spanish bank BBVA made Valora, which helps users calculate the best price to rent, sell, or buy a home, also implemented Baby Planner to understand better how starting a family will affect their financial situation. The Spanish bank is also planning to deploy a new app that uses biometric technology to automate payments. It enables users to book tables at restaurants, place orders from a smartphone, and leave the restaurant after the meal without asking for the bill or pay manually.
- **FeedZai – Citibank** invested in Feedzai, an advance data science company, which allows them to go through large amounts of data in a flash and alert customers of fraudulent activities in real time.
- **COiN – JP Morgan Chase** Bank launched its contract intelligence platform to cut down on loan-servicing slips in 2017. The M.L. technology used review 12,000 commercial credit agreements in seconds, saving about 360,000 human work hours per year.
- **Wells Fargo** – made an AI-driven chatbot on the Facebook Social Network Messenger platform in 2017. It delivers live information to help clients make better financial decisions, from how much they spent on food, through balance on accounts and bill payment dates.

- **RBS – The Royal Bank of Scotland** uses an automated lending method to approve commercial real estate loans in less than 45 minutes. This process would typically take days. The AI-driven bot launch is part of the bank's more extensive digital and innovation agenda. Bank has also adopted a cognitive chatbot, powered by IBM Watson Conversation, to answer customer queries.

Payment-As-A-Services

PaaS is a cloud-based method for commercial payments. Basically, it moves your cash into the cloud and helps you automate the back-office financial process from end to end. In enhancement to this, instead of charging customers fees based on the transaction, subscribers only pay a flat monthly rate (subscription).

Payment as a Services is easy to use and manage. It's incredibly scalable because it doesn't need to be downloaded or installed on devices to onboard an entire finance team or business. It is mainly helpful for growing start-ups, large companies with multiple subsidiaries, and remote teams.

What Is the Purpose of PaaS?

Payments-as-a-Service evolved out of the increasing recognition that today's financial system is under costly fees, different data sources, and paper-driven processes. PaaS eliminates most of the cost and complexity from the traditional payment process, speeds up collections, and releases up accounting teams to focus less on paperwork. More on high-value activities like forecasting (Figure 7.2).

With this solution, top managers gain the ability to analyse data and cash flow in real time and make actionable decisions about financial management that improve and support revenue growth.

The advantages are:

- 3.2x higher ROI than on traditional solutions
- Zero-fee digital payment options
- 50% lower costs to manage the end-to-end process
- No additional hardware costs and no need for lockbox services
- Certified payment processor PCI DSS included

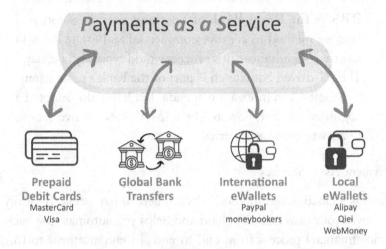

Payment and payouts under the PaaS model

Figure 7.2 PaaS model.

Benefits

There are loads of advantages of moving payments to a cloud-based solution. PaaS can provide notable advantages to the standard business environment. Key benefits are:

- **Easy Access:** Only web browser and an internet connection. Since all data is stored in the cloud, payment can be made from anywhere you want.
- **Fast Implementation:** It does not need to be installed on an individual computer/smart device. Once PaaS solutions set up an account, a company can start accepting payments immediately.
- **Improve Compliance:** PaaS solutions create notarized, **digital records** stored on the Blockchain to ensure that payments are secure and free of tampering.
- **Lower Costs:** Decrease monthly costs and enhance forecasting accuracy by switching from volume-based pricing to flat monthly rates.
- **Reduce Fees:** PaaS solutions offer several ways to reduce processing costs using the software. Companies can also

remove fees entirely using zero-cost payment rails, allowing clients to move money without paying any fees.

- **Shorter Monthly Close:** Use automation features to reconcile payments instantly, update reports in real time, and set up alerts that complete A/R tasks on your behalf.
- **Software Updates:** Expensive upgrades have become a thing of the past with PaaS. Instead, PaaS providers update the software on a regular basis (typically, every two weeks). These updates have new features and capabilities – often because clients like to have new functionalities.

How Does It Work?

PaaS software runs directly through your web browsers (MS Edge, Chrome, or Firefox – making it unnecessary to download or install the software).

Files that would typically be stored on a hard drive are stored online. Because the data is stored so that it is always accessible, you can log into a service and perform A/R tasks on any computer with an internet connection.

The essential financial records connected with your business are protected from nearly all of the catastrophes that could spell havoc for an organization if traditional storage methods are used:

- If someone breaks into business and steals a work computer, clients don't have to worry about them getting access to all of those financial records simultaneously. Financial records are safe because they were never stored on the stolen computer.
- Records are also safe from a fire, from a hard drive failure, from data loss, and other types of situations that company owners needed to worry about in the past. PaaS also takes the step to encrypt data to help keep it safe from malicious individuals trying to access it using operating systems, malware, viruses, and other means.

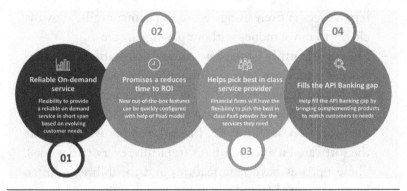

Figure 7.3 PaaS adoption, a cost-effective game changer.

To use a service, all is needed is to open a web browser (mobile application) of choice on a device, navigate to the official website, and log in using the credentials associated with the client's account. All of the client's account information will be available from your device, regardless of its size, operating system, technical specifications, or brand (Figure 7.3).

Impact (in a Nutshell)

- PaaS helps companies quickly configure new functionalities for better Return of Investment time.
- Companies can benefit from additional services such as AML screening.
- PaaS can help fill the banking API gap to seamlessly bridge complementary products and services with existing monolithic infrastructure.

Digital ID Is Becoming a Hidden Payments' Enabler for Retailers

In ever more challenging times, the financial and banking industry needs to maintain trust by finding a way to protect clients from the constant threat of payment fraud and theft. Consumers wish to limit physical contact during the COVID-19 pandemic, which has led to the popularity of contactless payments accelerated in multiple territories.

Digital identity (ID) is becoming a hidden payment enabler for retailers.

In the United States, one in five customers has made a contactless payment for the first time during the COVID-19 pandemic, according to research published in August by the National Retail Federation and Forrester.

At the 2019 Money 2020 conference, there was a universal call for a comprehensive digital ID to enable digital payments. A digital ID form would make cashless payment communications – intelligent, private, efficient, and secure. The conclusion was that without functioning digital ID, the payments revolution would stall.

In a connected world, consumers find themselves required to authenticate their ID daily, whether with financial institutions, government departments, healthcare, or retail stores.

Privacy regulations such as the General Data Protection Regulation (GDPR) have attempted to restore some trust, but the consumer industry still has a long way to go.

Unlocking a Bold New Ecosystem

Currently, authentication is fragmented and unwieldy. It requires a mix of documents, online login credentials, and digital wallets. This is frustrating for consumers and leads to the reuse of passwords and Personal Identify Numbers (PINs) that expose the consumer to fraud.

Mastercard Company concludes that there is a clear need for a verified ID that is accepted globally and across various digital touchpoints and doesn't involve aggregating more information in potentially vulnerable data stores. Instead, it gives the individual control over their ID data.

An integrated digital ID system would enable the payments industry to fight deception on a global scale. It would also meet the urgent need for a payment authentication way that consumers can access anytime, anywhere, and on any device. This joined-up approach is vital to ensure the digital transformation of the world.

Providing access to a singular, centralized digital ID will streamline the ID process and unlock new and improved consumer experiences during this digital transformation, particularly in the new breed of smart cities, where everything from housing, travel to payment systems will be connected to a digital user ID.

What Form Will the New Digital ID Take?

There are two significant challenges that financial providers need to overcome with a new ID solution: new users onboarding and securing the digital ID is fit with all activities (Figure 7.4).

Placing individual customers at the centre of their digital interactions will ensure trust and more progressive adoption of new technology payments and services. Yet, for this to be successful, the financial industry must adopt a simple, familiar, and easy-to-understand process, which from current monolithic architecture is very hard to make possible.

Fingerprint Biometrics as a Digital Identity

The use of fingerprint or face recognition authentication to unlock a phone is now profoundly established. As far back as 2016, 89% of users with compatible smartphones were using fingerprints to unlock their devices. The answer for frictionless client onboarding has been in front of our fingertips/face the whole time.

Financial industry providers can use fingerprint biometric sensors directly into their new breed of smart payment credit cards. A biometric card may be a new concept, but financial providers and

Knowledge Possession Biometric

Figure 7.4 Biometric identification.

retailers around the world are already using contactless technology in the payment process, so it is the next mandatory logical step.

Consumers are now carrying a card and using it for contactless payments. Plus, as it is seen, consumers are used to using their fingerprint as an authentication mechanism. Maybe biometric cards could be the catalyst for financial inclusion aspired by the World Bank, as they don't require expensive smartphones in emerging nations.

Building Trust with Biometrics

Continuous developments in financial regulation mean that secure authentication is crucial. Under PSD2 (Second Payment Services Directive) European regulation, all payment transactions require Strong Customer Authentication (SCA) to validate users' point of transaction to reduce fraud and increase customer security.

SCA demands two factors' authentication for every transaction above the contactless limit. While one is something you have like a smart card, the second can be a fingerprint, OTP, face recognition, or push message. Using a fingerprint/face recognition means that it can be used on multiple platforms.

Biometrics is playing a crucial role in digital ID, significantly limiting endangerment to potential fraud and criminality. A biometric sensor's addition onto a credit/debit card creates a safe "chain of trust" that indelibly unites the user and card.

Moreover, a digital ID can be extended far beyond payments and used as a unique identifier in access, government ID, and even IoT devices.

Securing the Future of the Financial Industry

While the world is becoming eternally cashless, commentators and analysts agree that the payments revolution will stall without a fully functioning digital ID.

If we fix digital identity, we fix payments.

Tony McLaughlin, Emerging Payments and Business
Development at Citi Bank

Payments and the consumer industry need a user-centric digital ID owned and managed by the individual to unlock the full advantages of a transformative digital payment ecosystem.

Using fingerprint or face recognition biometrics as a digital ID in a payment card will transform the way people verify transactions. This integration would allow consumers to confirm ID wherever they are and across every transaction. It will change the way of digital ID as we know it.

Digital communications should be intelligent, privacy-enhancing, efficient, and secure. It is time to place people at the heart of their digital interactions globally.

In a Nutshell

What Will Be Good to Take from This Chapter?

Machine Learning   Financial institutions, especially banks, sit on a mountain of customer data such as social background, purchase history (products purchased and amounts paid), and household income. Thanks to the model shift in sales from reactive to proactive, this data can be leveraged, moving from instinct-driven to insight- and data-driven.

The algorithm uses data and feedback from humans to learn the relevance of provided inputs to a presented output (e.g., how the inputs "air pollution" and "interest rates" predict real estate prices). There are many different algorithms for the many different business cases. And algorithms are used in different types of analytics, which can be classified based on their complexity or application domains. It will be good that we can explain:

- Linear Regression
- Logistic Regression
- Decision Tree
- Native Bayes
- Neural Network

Payment-as-a-Services Payments-as-a-Service evolved from the growing realization that today's financial system suffers from costly fees, disparate data sources, and paper-driven processes. PaaS eliminates much of the cost and complexity of the traditional payments process, speeds up collections, and frees up accounting teams to spend less time on paperwork and more time on higher value activities like forecasting.

This solution gives top managers the ability to analyse data and cash flow in real time and make actionable financial management decisions that improve and support revenue growth.

The benefits are:

- 3.2x higher ROI than traditional solutions
- Fee-free digital payment options
- 50% lower cost to manage the end-to-end process
- No additional hardware costs and no need for lockbox services
- Certified payment processor PCI DSS included

8
BLOCKCHAIN BANK ARCHITECTURE

The difference between good and bad architecture is the time you spend on it

— David Chipperfield

What Could Be Exactly Blockchain in the Banking Context?

As we mentioned in previous chapters, Blockchain technology is a distributed, open ledger that permanently records transactions between two parties. Both involved parties can share a digital ledger across a network without needing intermediaries or centralized authority. That's why transactions processed through Blockchain network are faster.

What are the prerequisites that need to be discussed before building a Blockchain bank? One of the first should be Know Your Customer (KYC). This is the process of recognizing and confirming clients' identities. The condition is also used to find the bank and anti-money laundering (AML) laws that rule these activities.

AML software is a cast of computer programs employed by financial institutions to examine customer data and catch up with suspicious transactions (Figure 8.1).

Know Your Customer Steps

Once a client is registered on the website, the client should enter cryptocurrency wallet details and the amount that they want to contribute.

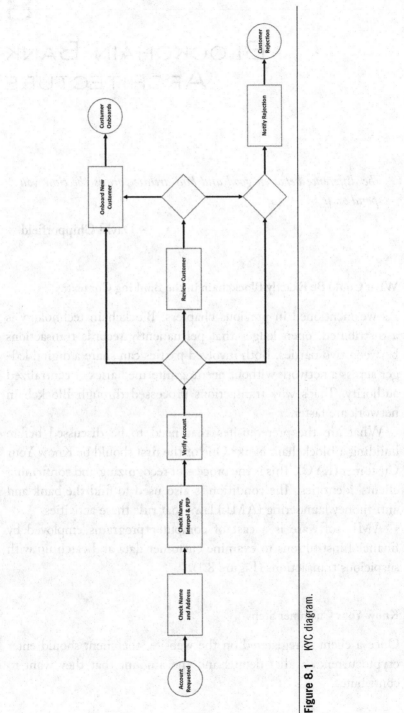

Figure 8.1 KYC diagram.

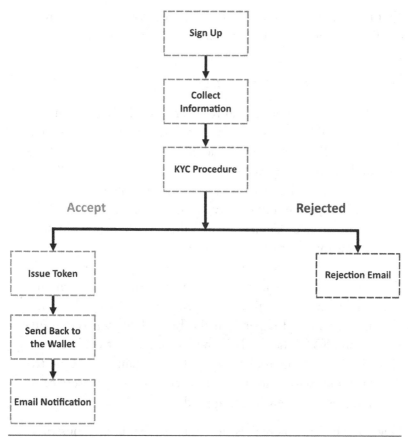

Figure 8.2 KYC steps.

The first thing that needs to be confirmed is KYC. A client can use various documents, like a passport, identity card, or driver's license. The next step is to verify residency. Again, multiple options can be used, e.g., utility bill, bank statement, or official government letter that shows client's address (Figure 8.2).

KYC Push

Not everyone is happy about KYC, and people don't like those traditional methods, mostly when it comes to cryptocurrencies, where transfers are anonymous or hidden. However, banks need to be sure that the sources of all funds raised are authorized and legal.

No funds can come from people under penalty, criminal sources, or red-flagged organizations with terroristic connections.

What Are the Benefits of Blockchain KYC?

- The first is that a general KYC and AML Blockchain registry could be generated, which could be used by different banks. Blockchain could accelerate the onboard working and dramatically reduce the costs of KYC compliance. Every time that a bank is signing up a new customer, a bank employee could be given access to the customer's KYC information except having to walk the customer through all of this information every single time.
- Another key profit is that a KYC and AML registry could be generated for intra-bank use. It means that when customers are using different services provided by the bank, the bank could depend on the Blockchain registry to fulfil the KYC and AML compliance without dealing with all of the recognitions over and over again every time the customer wishes to utilize a recently developed service or purchase of a new banking product.

One of the fundamental issues that the banking industry confronts today is the expansion in misrepresentation and digital cyber assaults. Presently, the more significant part of managing account frameworks is based on a centralized database, making them more vulnerable to cyber assaults as all data is put away locally in one place. Additionally, numerous banking frameworks are monolithic, obsolete, and helpless against new types of digital assaults.

By creating new managing account frameworks over Blockchain innovation, the possibility for extortion and information stealing can be significantly decreased as the distributed record innovation secures records; it stores, scrambles, and checks every piece of information in exchange.

Accordingly, should any information breach or false change happen, it would be made promptly evident to all parties who have consented to get to the exchange information on the record.

Consistence and KYC methods have become progressively imperative in managing an account industry as regulators keep an eye on banks' identity, working with to stay away from potential illegal tax avoidance or fear-based oppressor financing. As per a Thomson Reuters' Survey, the money associated organizations spend by and large $60 million on KYC and client due.

Regulators need better access to banks' client customer bases and transaction histories, while banks need to consent to the regulator's desires to keep away from administrative fines no matter what. By creating consistent steps and KYC forms over piece chain innovation, banks can decrease operational expenses in these offices and increment the productivity of consistent forms and build up a nearer relationship with the money-related regulators.

Since Blockchains can store any computerized data, including computer code that can be executed once at least two individuals enter their keys, Blockchains allow us to have smart contracts. The code could be modified to make agreements or perform money-related exchanges once a specific arrangement of criteria has been met – transport of items could flag a receipt to be paid, for instance.

In the proposed system, the traditional banking architecture which consists of a centralized database will be removed. The data will be distributed over the Blockchain network, which will make the banking systems decentralized. This will make the data more secure, and also it will remove the power centralization. The transactions over the Blockchain will be in the form of encrypted tokens, which each node on the Blockchain will verify. To make any transaction legitimate, the nodes of the Blockchain will have to prove the processing it has done to verify the transaction. That proof will be taken in terms of the amount of processing done.

The above transactional system has two advantages (Figure 8.3).

1) It will make transactions faster by removing the intermediate processes used in everyday transactions.
2) It will become almost impossible for an individual to break the system as it requires an enormous amount of computing power that no one has.

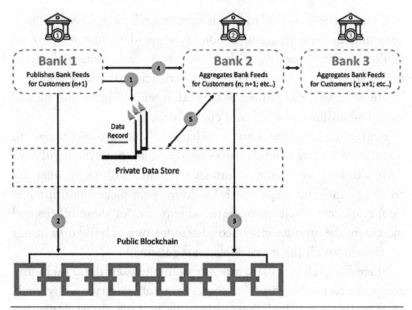

Figure 8.3   Blockchain bank architecture.

The architecture diagram shows the implementation of the banking system on the Blockchain architecture.

- **Public Blockchain** –The public Blockchain is the distributed system that will contain the decentralized banking data.
- **Private Data Store:** Some data cannot be decentralized, so those data should be stored in a centralized system.

The bank will publish a feed of the customers over the Blockchain, aggregated by intuit. Since the feeds will be decentralized, it won't be easy to breach them.

In the existing system, the banks have centralized databases in which every customer's data is stored. The transaction process is too slow. It has to be authorized by at least four different bodies, says: the payment portal, the gateway, the receiving account, the sending account, etc. This makes the transaction slower. Moreover, it is easier to control even one of these bodies and take control of the whole transaction system. It is expensive for the banks to maintain such large databases, transaction systems security, etc.

In the future, the privacy issue in Blockchain can be removed. This theory can actually be implemented in the existing banking systems, making the banking systems more secure and fast and helping the banks and the government eradicate the black money problem.

FinTech is one of the emerging fields in today's world, and investors are risking big on it. FinTech adoption is growing at an exponential rate in India, and it is second only after China in the place of adoption. Part of it is the Blockchain boom, and the financial sector has been benefited from it the most.

The architecture of a traditional banking system (Figure 8.4).

But after the implementation of Blockchain, it will look something like (Figure 8.5).

When implemented, Blockchain can save billions in cash which the banks usually spend in different cases. For instance, KYC, each bank maintains its own database of customers and spends anywhere around 60 million to 500 million USD in the process.

Since the whole system is decentralized and a single entity doesn't control the deposits, the systems cannot go bankrupt.

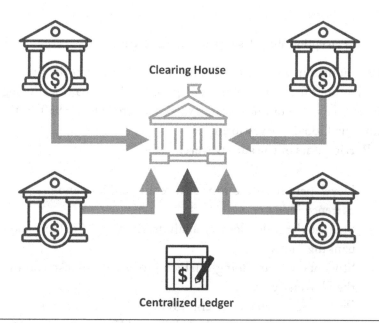

Figure 8.4   Traditional banking – centralized ledger.

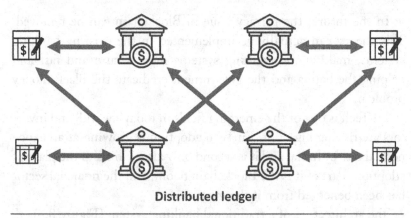

Distributed ledger

Figure 8.5 Blockchain banking – distributed ledger.

Smart contracts can be implemented, which will eliminate the need for any middleman. One prominent example of it is in microgrids. Using Blockchain in microgrids will change the way we pay for and use electricity. One more application of it can be in developing nations to decrease corruption from the real estate market and eliminate the so-called Block Economy.

Public Perception about Using Blockchain Technology

After numerous surveys on Blockchain technology in the financial and banking field, the public perception of Blockchain technology is also taken. The online research was conducted among 950 participants via Google survey form.

Participants of the survey were:

- Developers from the FinTech, retail banking, and cryptocurrency industry
- Stakeholders, developers, analysts from the FinTech and banking industry
- Students who are doing FinTech projects on Blockchain in the University
- Professors, teachers of technical fields from universities
- Researchers who know Blockchain and FinTech technology

The following questions were asked:

1. What kind of impact will Blockchain have on the financial Industry?
2. In your opinion, what is the core of the following banking features of Blockchain technology?
3. In which industry Blockchain technology can create the most value in?
4. At what level you trust Blockchain technology?
5. What are the challenges for the evolution of Blockchain technology?

Following conclusions can be drawn from the above questionnaire:

The majority of people think that Blockchain technology will significantly impact the FinTech and banking community. Since Blockchain technology is a decentralized system, it provides enough security for business data and also it protects the data from intruders (Figure 8.6).

Survey participants think that distributed systems like smart contracts are the most significant features of Blockchain technology (Figure 8.7).

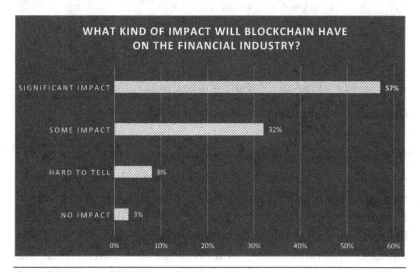

Figure 8.6 Impact of Blockchain on financial industry.

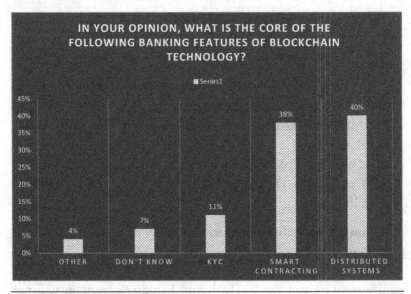

Figure 8.7   Core of the banking features.

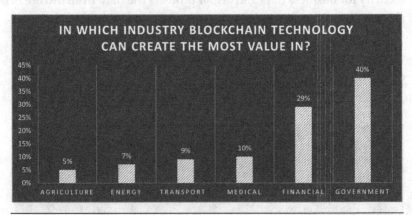

Figure 8.8   Blockchain industries.

Survey participants also feel that the government sector is the most promising sector where Blockchain technology can be implemented. We all know that government has most sensitive data of their citizens including personal identification number, health data, biometric information, transportation information (such as vehicle number, registration date), location information, etc. Hence, the data need to be protected (Figure 8.8).

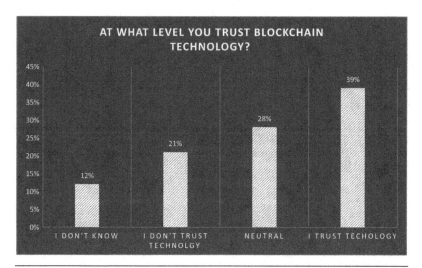

Figure 8.9 Blockchain trust.

Welcoming the new technology – Blockchain is the need of the hour. From the survey we have conducted, almost 40% people have high confidence in this technology. There is still a need for awareness which is required. We still have a shortage of professionals in Blockchain technology. There is a need for new courses on Blockchain technology to understand the core concepts, how it works, and the challenges that Blockchain technology faces today (Figure 8.9).

Speaking about the challenges of FinTech Blockchain technology, around 41% of people have admitted that we lack an understanding of Blockchain technology. On the parallel line, 19% of people also think that market readiness must accept Blockchain technology. No doubt, in the upcoming years Blockchain will be one of the most trustworthy, acceptable, and secure technologies globally (Figure 8.10).

Banking-as-a-Service

"Banking as a Service (BaaS) is the provision of complete banking processes (such as loans, payments or deposit accounts) as a service using the secure and regulated infrastructure of an existing licensed bank with modern API-driven platforms"

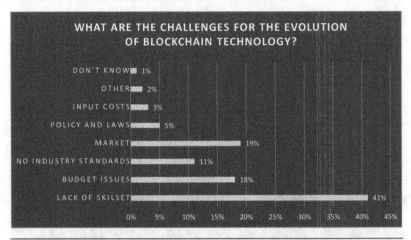

Figure 8.10 Blockchain challenges.

Digital transformation is opening data to enable more comprehensive transparency and better consumer experiences. New technologies are opening up monolithic systems to emerging start-ups and third parties and putting data directly in the hands of consumers.

How Banking-as-a-Service (BaaS) Works?

Banking-as-a-Service (BaaS) platforms have surfaced as a crucial element of open banking in banking and financial services. Companies provide more financial transparency options for account holders by opening their APIs for third parties to develop new services.

FinTechs and digital banks have been exceeding on necessary organizations in the banking-business game and disrupting traditional industry models – but by moving into the BaaS, legacy banks can turn this emerging threat into an opportunity.

What Will Be Banking-as-a-Service?

BaaS is an end-to-end consumer model that allows digital banks and other FinTech companies (third parties) to connect with systems directly via APIs and build banking offerings on top of the providers' regulated infrastructure.

Tech-savvy legacy companies can parry off the emerging threat of FinTech by migrating into the BaaS to share their data and infrastructure. Very soon, access to this level of information will become crucial for digital customers, and banks that are transforming now will be ahead of the competition and likely rewarded with high consumer demand.

The BaaS model begins with a FinTech, digital bank, or other third-party providers (TPPs) paying a small fee to access the platform, the banks.

It opens its APIs to the TPP and grants access to the systems and information necessary to build new financial products or offer white-label services.

The two main monetization strategies for BaaS include charging clients a monthly fee for access to the BaaS platform or charging for each service used.

Several countries have already begun introducing digital open banking laws, showing that the banking services business moves towards an era where shared infrastructure and data will become standard consumers' expectations.

In the United Kingdom, the new income potential generated through open banking-enabled SME business and retail client propositions was £500 million ($700 million) in 2018, per PwC – and Insider Intelligence expects that to grow at a 25% compound annual growth rate to reach £1.9 billion ($2 billion) by 2024.

Beyond adding a new revenue income, developing a BaaS solution also enables legacy banks to establish relationships, form partnerships with rising FinTech, and keep themselves ahead of the courses that will inevitably follow once BaaS and open banking become mainstream.

What Does BaaS Contain?

BaaS stack shows a standard view of the banking stack, breaking down on modules and practices that make banking ecosystem. BaaS allows any brand to embed financial services into its consumer experience by picking and choosing services offered by providers and license holders in a modular fashion.

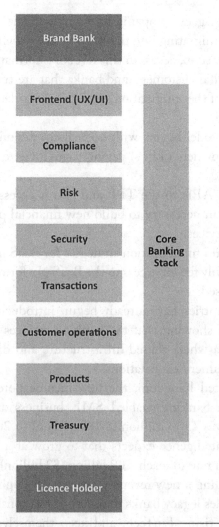

Figure 8.11 BaaS stack.

Brands encapsulate financial services in their client offering. Providers offer modular business capabilities to brands to embed finance. License holders partner with providers and give access to financial products.

BaaS enables (Figure 8.11):

- Collaboration BaaS is about working together. It requires collaboration from multiple providers.

- BaaS allows companies to design and build new projects far faster than they could on their own.
- BaaS allows services to be delivered by an ecosystem of specialized providers rather than one single company.

Most of the brands have different commercial intentions but have all introduced card-based payment services to their clients through BaaS providers. The BaaS providers default to offer functional, clean APIs with developer-first procedures to sales and marketing. These companies can quickly get payment services to market and improve the commercials as they scale and apply for new licenses.

Traditional banking brands are license holders and sometimes providers of white-label or co-branded cards. However, some of these brands look at BaaS providers as payment gateway processors to help them take their balance sheet and market capabilities. Few established banking brands have the abilities and improved API-driven platforms needed to attract non-finance digital brands as clients.

New providers empower brands to deliver embedded finance and better products faster.

Three BaaS providers stand out in the United States of America:

- Galileo has a set of APIs and partnerships with more than 20 banks and can offer prepaid, debit, and investing capabilities.
- Marqeta concentrates on providing payments and debit card programs through its API services. It partners with a couple of small U.S. banks. Marqeta provides brands digital, real-time experiences.
- Synapse provides payments, lending, deposits, and investment products. It is simpler and more out of the box than Marqeta's offerings to date but less compliant and robust for brands as they scale.

European BaaS are more diverse:

- Railsbank operates with payments processors, partner banks, and FinTech to enable the rapid creation of financial products.

- Solaris Bank powers many of Germany's largest FinTech players. It is offering API capabilities in the BaaS stack. Unusually, it has its license and has correctly coupled the regulated activities with precise APIs.
- GPS was the payments processor behind the United Kingdom's early FinTech leaders, including Revolut, Curve, and Monzo. With a broad geographic footprint, GPS can offer payment services and work with local partners to access card schemes.

Lots of mostly smaller banks now offer their capabilities as a service, particularly in the United States. Smaller U.S. banks are using BaaS partnerships to attract deposits.

BaaS needs improved, modular API-driven banking software platforms, but BaaS and banking software are not the same.

Banking Software Will Become Modular

Improved banking software platforms share the same cloud-based, modular, API-driven strategy to architecture fundamental to delivering BaaS. Enhanced banking software providers represent a significant advance on monolithic legacy banking software. These software providers sell primarily to bank license holders and require a level of balance or expertise to implement or upgrade their platforms over time. By providing bank software as a service, modernized banking platforms bring license owners increased speed to market, the pace of change, and expertise to configure and personalize financial products.

Banking Software Platforms Are Not BaaS

To manage a banking software platform, you must either be a bank or have some lending or payment license level and a relationship with a bank.

BaaS is sometimes confused with banking software delivery as a Service, but they are not the same thing (Figure 8.12).

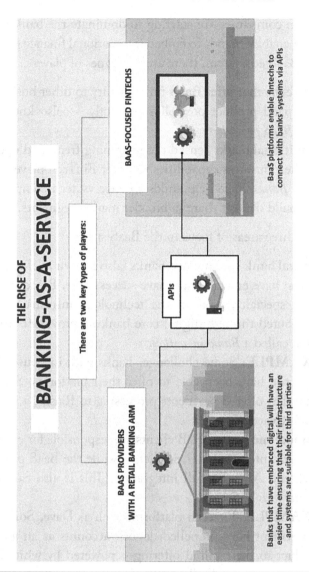

THE RISE OF

BANKING-AS-A-SERVICE

There are two key types of players:

BAAS-FOCUSED FINTECHS

BaaS platforms enable fintechs to connect with banks' systems via APIs

APIs

BAAS PROVIDERS WITH A RETAIL BANKING ARM

Banks that have embraced digital will have an easier time ensuring that their infrastructure and systems are suitable for third parties

Figure 8.12 BaaS APIs.

Software as a Service is using centrally hosted delivery and licensing on a subscription basis. Software as a Service is referred to as on-demand software. Modernized core banking platforms are usually delivered as a service. Any brand can partner with a BaaS provider to provide bank-like services.

Platform companies are starting to dominate the banking business. BaaS enables brands to embed international finance payments. In the BaaS ecosystem, there are two types of players:

1. **FinTechs that offer BaaS** functionality to other businesses to expand product portfolios promptly — also known as *Pure BaaS providers*;
2. **Traditional banks** sensed the upcoming trend and opened up their business capabilities to other FinTech players. For these *Banks*, *BaaS* help builds a hedge vs. tech competition and build deposit share in broader market segments.

There are three areas of focus in the BaaS space:

- **Virtual bank** – Many neo banks (also known as challenger banks) have emerged and have success here. Based on this BaaS specialty, most of the technology infrastructure is distributed through digital core banking providers. This is often called a *Bank-in-a-Box*.
  EXAMPLE: Some challenger banks with in-house BaaS expertise have been able to offer their platforms; Starling Bank has offered payments processing to Raisin, a savings marketplace.
- **Standalone bank** – B2B clients are responsible for creating the user interface for their clients, while the BaaS provider back-end APIs for full integration. This is also called an *Accounts-as-a-Service*.
  EXAMPLE: FinTech platforms such as Dave, SoFi, and Betterment have propelled deposit accounts as an add-on product to their initial offering – powered by white-label BaaS providers or bank partnerships.
- **Upgrading the outdated infrastructure of legacy banks** — Bank institutions have grown in size over the last decades from multiple acquisitions of smaller banks. The long-term effects have been outdated legacy systems, leaving technology gaps within the larger entity.
  EXAMPLE: BaaS providers migrate older tech stacks into efficient banking cores that are all cloud-connected and hosted by TPPs.

Blockchain Payment Software

Blockchain payment software allows us to facilitate, process, and verify transactions conducted on a Blockchain or distributed ledger system. These tools are designed for individuals, businesses, or financial institutions and have different features depending on the business case. Blockchain payment solutions enable customers to quickly and securely transact on a Blockchain across all business applications.

Some tools enable creating smart contracts or protocols used to facilitate, verify, and execute transactions. These contracts are fully documented and provide a high level of traceability. Forms of payment are quicker than traditional tools since they do not require a financial institution and transactions can be completed across country borders with ease.

Some of the less complex and customizable Blockchain payment products offer a product similar to payment gateways. The only difference here is the Blockchain's documentation and traceability and the simplified international payment process. These tools are not used for trading anything except cryptocurrencies such as bitcoin or Ethereum.

To qualify for inclusion in the Blockchain Payments category, a product must:

- Be made on a Blockchain-based platform or distributed ledger technology system.
- Process and authenticate transactions on a Blockchain.
- Facilitate transactions without a financial institution, bank, or another third-party verification service.

The biggest companies in the world are already deep inside Blockchain platform software. Companies like:

Mastercard Blockchain: promotes new business opportunities for the digital transfer of value by allowing businesses and financial institutions to transact on a DLT. Technology can power multiple business cases and can reduce time, cost, and risk out of financial flows.

Samsung Blockchain Platform: a gateway to development and monetization, providing talented developers and designers with global distribution opportunities, end-to-end support, and easy-to-get resources. The ultimate goal is to give developers the tools and support they need to build, market, and monetize products on Samsung's on-device app store. The store is deployed in more than 90 countries, with over 202 million monthly active users and 1.6 billion monthly app downloads.

Civic Reusable KYC: a Blockchain-based tool for organizations that require more than essential account verification services.

IBM Blockchain Platform: Blockchain is a shared, immutable ledger for recording the history of transactions. It fosters a new generation of transactional applications that establish trust, accountability, and transparency from contracts to agreements to payments.

Azure Blockchain Workbench: presents a rapid, low-cost, low-risk, and fail-fast platform for companies to collaborate on by experimenting with new business processes, and it's all backed by a cloud platform with the biggest compliance portfolio in the industry.

Amazon Managed Blockchain: a fully managed service that makes it easy to create and manage scalable Blockchain networks using the popular open-source frameworks Ethereum and Hyperledger Fabric.

Blockchain-based Payment System

A payment processor is a company that conducts payments on behalf of individuals to companies. A payment processor company processes a credit/debit card transaction, and a company gets paid minus the fee, which is divided between the credit card company and the processor. The main benefit is comfort, security, and the capability to deliver money anywhere in the world with one single form of payment.

Payment processor companies are reluctant to accept payments with cryptocurrencies such as bitcoin because of rapid market fluctuations and the risk of money laundering and fraud.

But, it is possible to accept bitcoin (or any other cryptocurrency) for products offered online and at POS terminals. Multiple companies provide bitcoin payment processing securely and automatically convert the payment into the currency of choice.

The ability to make business payments or offer Blockchain payment solutions for companies is the reality of few technology start-ups usually enclosed in the FinTech space. Multiple business models provide these services, including investment, currency conversion, and payment between fiat currencies and cryptocurrencies.

Bitcoin payments are becoming more popular with shops and online retailers by the thousands already accepting it and other cryptocurrencies as part of their payment channels.

The technologies are vast, and they will encounter solutions in varying levels of robustness and adoption. For example, it is possible to start accepting bitcoin payments right away on e-commerce webshops by signing up with a few companies.

In Japan, there are multiple companies, including national retailers, already accepting crypto payments in their stores. The adoption of bitcoin payment technology is a bit slower in other markets, but it is still prevalent in Germany and the United States.

Bitcoin for B2B payments would be the most significant solution to problems of KYC, contracts, and currency conversion for international payments.

With mass bitcoin adoption for B2B payments, KYC would be minimized because all of the processes would be as simple as accessing a Blockchain-based KYC system connected to a company-owned IBAN. The contract would be in the Blockchain, eliminating hard checks and mailing. The money would not take time to be transferred to the home account: it would be available instantly, reducing risk and the need for factoring.

In a Nutshell

What Will Be Good to Take from This Chapter?

Banking-as-a-Service "Banking as a Service (BaaS) is the provision of complete banking processes (such as loans, payments or deposit accounts) as a service using the secure and regulated infrastructure of an existing licensed bank with modern API-driven platforms"

Digital transformation is opening up data to enable broader transparency and better customer experiences. New technologies are opening up monolithic systems to emerging start-ups and TPPs, putting data directly into the hands of consumers.

How does Banking-as-a-Service (BaaS) work?

BaaS platforms have emerged as a critical element of open banking in banking and financial services. Companies provide more financial transparency options for account holders by opening their APIs to third parties to develop new services.

FinTechs and digital banks have overridden necessary organizations in the banking industry and are disrupting traditional industry models – but by entering BaaS, long-established banks can turn this emerging threat into an opportunity.

Blockchain Payment Software   Blockchain payment software allows us to facilitate, process, and verify transactions conducted on a Blockchain or distributed ledger system. These tools are designed for individuals, businesses, or financial institutions and have different features depending on the business case. Blockchain payment solutions enable customers to quickly and securely transact on a Blockchain across all business applications.

Some tools enable the creation of smart contracts or protocols that are used to facilitate, verify, and execute transactions. These contracts are fully documented and provide a high level of traceability. Forms of payment are faster than traditional tools as they do not require a financial institution and transactions can be easily executed across countries.

Some of the less complex and customizable Blockchain payment products offer a similar product to payment gateways. The only difference here is the documentation and traceability of the Blockchain and the simplified international payment process. These tools are not used for trading anything other than cryptocurrencies like bitcoin or Ethereum.

9

COMMON FEATURES
Banking of Tomorrow

When I'm working on a problem, I never think about its beauty, I just think about how to solve the problem, but when I finish it, if the solution is not beautiful, I know it's wrong.

– Richard Buckminster Fuller

Transactions of Tomorrow

Money Is Improving Itself – Banks Must Do the Same

Banks Look to Save Their Place in Tomorrow's Economy

Three key questions banks and FinTech companies should embrace to be competitive in tomorrow's digital economy.

Technology and progress are revolutionizing how we live. One of the visible effects of this is a technological change in how we exchange money for day-to-day goods and services with each other. It has rapidly transformed into real-time contactless payments and mobile-based banking from cash.

It may only be a matter of a year before societies are virtually cashless and transacting in digital and virtual currencies of one form or another.

These changes in how we conduct financial transactions raise questions for today's banks. Where will they operate and what roles will they take in a payment ecosystem as demand for branch services and cash-dispensing ATMs falls?

DOI: 10.1201/9781003198178-9

COVID-19 Accelerates Digital Banking Transformation

No doubt, the COVID-19 pandemic that quaked the world will prove to be a significant accelerator. During the pandemic outbreak, cash usage declined, as did the use of ATMs. There was a massive spike in online and mobile banking. Online and cashless PoS transactions accounted for around 60% of total transactions, up from 20% to 30% pre-COVID-19.

Mass digital adoption is not easy to achieve; banks need to ensure that customers' experience remains seamless and easy if they want those new clients to become digital and use mobile and online as a primary channel.

The financial transaction arena is one of the busiest areas of change and innovation. The biggest tech companies grow their payment services. New FinTech emerges offering mobile digital wallets or flexible "buy now pay later" options that have seen increased take-up recently in some geographies. Traditional banks and financial institutions risk seeing their market shares being diminished around them and losing their authority in the customer relationship.

Banks need to strike back! Significant investment needs to be directed into their digital offerings and develop API strategies that link their banking platforms to payment services (A European PSD2, directive demand that banks open their APIs to the third parties), data providers, online media, and other services (Figure 9.1).

Capgemini forecasts that between 2018 and 2021, the compound annual growth rate for all digital transaction volume will be about 13 percent, rising to 876 billion from the current estimate for 2018 of 598 billion. Growth is fastest in developing markets, and mature countries are moving at a much slower rate.

Capgemini

Tech Partnerships Will Be Essential

For many companies who need to accelerate their digital agendas quickly, an essential aspect will be to make acquisitions of potential

A new, digital world

Total digital transaction volume is on track to grow 46%
over the next four years

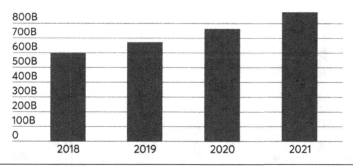

Figure 9.1   Volume of new digital world transactions.

technology payment companies and forge partnerships and joint ventures. A good example will be Mastercard and Revolut.

In the time of the COVID-19 pandemic, this is set to become even faster. Banks have woken up that creating digital payment products and services is not just nice to have but a must have. While they have developed solid digital offerings of their own, traditional banks' systems are not sufficiently "straight through", and the go-to-market is not where they would like it. Making precise collaborative partnerships could be a key determinant of success.

Also, banks are likely to need to make more partnerships with some of the network giants such as Visa, Mastercard, American Express, or maybe PayPal in the future? Through acquisitions of their own, these businesses are already curating a set of FinTech, data, and payments capabilities that banks can leverage to access new capabilities without integrating complexity.

Another extension area is likely to provide retailer and payment services to small and medium businesses. Most of the banks have focused on providing services to large corporates. Still, as more small businesses embrace digital services and PoS systems, there is a significant opportunity to offer them treasury management and liquidity services and become real business partners in raising their growth and development. The solution for success here will

Top of the world

Russia and China lead the world in adopting digital payments

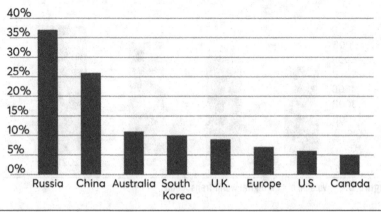

Figure 9.2 Digital payment adoptions.

be ensuring the banks can deliver simple, easy-to-use, cloud-based digitized solutions (Figure 9.2).

> Russia leads the world in its digital transaction growth rate, with a recent 37 percent surge in total volume from regulations capping cash payments, new electronic payment options and the National Payment Card System's introduction, according to Capgemini. China also is seeing strong conversion from cash and other payment forms to digital payments, followed by Australia, which saw a recent 11 percent increase in digital payments transaction volume.
>
> Capgemini

Rise of Digital Currencies

Central bank digital currencies (CBDS) have started to move onto the agenda actively. By making digital currencies official, they could change the face of how money works. China is at the forefront of this, and in the European Union, the French central bank has also begun a proposal to develop a coin. The Central Bank of England recently published an extensive white paper on the CBDC subject.

It may only be a matter of a time (next couple of years) before seeing a big CBDC launched, with the smart money (or crypto money) likely to be in China (digital yuan). Being a code, a CBDC will be like programmable money with many business cases and services possible beyond a digital replication of cash. Digital Coin backed by the central bank can support and enhance countries' financial stability. In China, a digital yuan could be a significant step towards exterminating the grey and black economy.

CBDCs may also be necessary for the unbanked, giving them access to payment services and enabling people to transfer or "send money home" to developing economies. Presently, cross-border payments carry high remittance fees. Central banks could program money in aid programs such that it is only usable for the aid or other purposes for which it has been remitted.

Add to these significant drives like the European Payments Initiative (EPI), which is created to produce an innovative and integrated payments system across the European Union, and projects like X-Pay in Germany to balance digital payment offerings.

Data and Software Are the Heart of the Digital Transformation Economy

The other significant trend to emerge is the development of new digital technologies using data as a service. Increasingly, transaction data is being connected with additional key-value-pair information to support new digital service development and capitalizing on open banking – this is the thesis behind some of the largest payments deals announced recently by credit card providers, cloud services, and online platforms.

The software traverses traditional or retail/wholesale payments and connects real economy business cases baked up with financial services. These are the new business models owned by Klarna (Swedish FinTech company), Gojek (an Indonesian Uber-like company), PayPal, and many others.

Banks will compete through their data assets, resulting in material acquisitions of data capabilities to retain relevance in the digital economy. Skills, mindset, and culture will be as much a driver for M & M&A as will the ability to novate legacy technology and corporate debt with scale instead of incremental change from within, which remain outpaced by the market.

Every bank needs to move decisively to find its niche. In a low-interest-rate environment offering thin margins, they need to find ways of maintaining profitability from payment services while preventing themselves from being disintermediated by new agile service providers. Operational resilience and strong cybersecurity will be essential, as operating efficiency will likely be outsourced with increasing non-core services.

Savings of Tomorrow

The banking industry continues to be enclosed by darkness, and there is no clear path for success. While banks are reinventing their organizational structure to be more innovative and develop more digital-oriented products, small start-up companies are overtaking the financial stage. These start-ups are reducing the complexity and increasing the transparency of banking services for the mobile-first generation. Some of them are focused on improving savings and managing personal finance. They are probably going to redefine the future of how consumers govern with their incomes.

Clients that save money have the pleasure of having money in their bank account to feel safe or make a big purchase with the saved money. But to highlight, more than 70% of adults have less than 1000$ savings. The formula of savings is straightforward:

Savings = Income − expenses.

Therefore it is necessary to have a certain amount of income that is enough to cover living and everyday expenses and have some part of the income amount left for saving purposes. In most cases, people cannot just afford savings because they don't have multiple incomes. But sometimes they can afford savings, but they waste on spending money rather than saving. 57% of consumers end up spending more than they have planned (Figure 9.3).

Though being very important, the process of savings is tedious. Not surprisingly, there are lots of exciting gamification that are

www.capablewealth.com

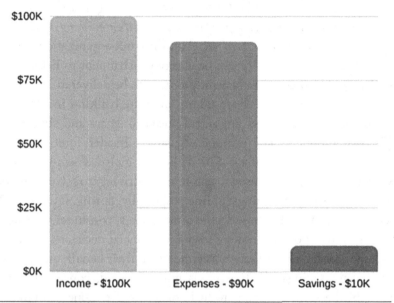

Figure 9.3 Incomes vs. expenses.

making people spend. Marketing, packaging, discounts are very few of so many nice things causing you to make easier spending decisions. But we don't have many of these types of cool things to encourage people to save easily. And in this digital-first era, when everything is digital, savings still are happening in the branches of traditional banks. 53% of customers visit their bank branch at least once per month to make deposits or withdrawals. Why can't we "ask" a bank to put $20 on our saving account, but instead, we have to visit the bank?

Machine learning and artificial intelligence are taking over everyday tasks, and banking is not expelled from this process. In banking, AI and M.L. help you faster analyse the vast amount of data to help you make better, quicker, and reliable decisions. Whether it is a saving decision, investment decision, or a spending one, they are just assisting you in making a smarter one with powered high-speed data analytic algorithms. These trends are very fast. Good examples can be *a new generation of mobile applications make transferring money less expensive and instant; advanced standalone companies use*

.

complex algorithms to create and manage portfolios, consumers use online platforms to borrow and lend money without visiting a middle man (a bank), consumers give saving platforms access to their banking accounts to make savings on behalf of them, etc. And the bank organizations seem to realize that technology can be a better instrument to help them understand how consumers want products to be delivered.

As technology and AI are taking over the banking industry in various ways, consumers are getting used to smart and simplified FinTech products. For example, Business Insider forecasts that Robo-advisors will manage $10 trillion in global assets by 2022. Consumers are getting used to autonomous driving cars. Consumers are used to buying things online, especially during COVID-19 lockdowns. With all these techie atmospheres, consumers are step by step relying on technology to make financial decisions on their behalf, manage their assets, investments on their behalf, and in the end, make savings. As the trends move forward, consumers will probably be more adaptable to technology to save money and manage their finances.

The future of the savings is something like this: Mobile solutions powered by Artificial Intelligence, Machine Learning, Automation, Robotization, Virtual Assistants making smarter and better-saving decisions on behalf of individuals. With the limited chance for significant improvement in producing simplified savings products and increasing pressures from mobile-first-oriented consumers, it is clear that doing more of the same and designing another savings product is not a good strategy. Most banks need to resolve what model they need to work towards and commit the required financial and human resources to make the transition better and painless.

Money-saving Isn't Easy

Whenever the pay cheque comes, most average consumers think, "This is the month I'm going to start saving". The intentions are always good, but then, well, same old story.

There's a nice pair of shoes, or a book, or a new gadget... And then, poof, before most consumers even really realize how much they're spending, the money is gone.

What's the best way to save? What's the first step? Tomorrow's technology, some of which is available today, can help. Many online services offer an online account with a free debit card.

Some of them are also a great help in saving money. And they do it without the consumer really realizing it.

Spending Budget?

With online services like that, consumers can add money to account whenever they want and then pay for things directly with that debit card or phone. That means that it is possible to make a budget by only adding a certain amount of money each month and not allowing to spend beyond that.

Lower Fees for Spend Money Worldwide

With an online card like, e.g., Revolut, consumers can spend worldwide (in 150 different countries) while avoiding high fees. It is also possible to exchange currencies and open multi-currency accounts on a client's mobile phone – 30 of them at once.

Vaults Spare Change Savings

Every time a consumer purchases Revolut, it will automatically round up the transaction to the nearest higher whole number and send the difference to personal Vault. The money will stay, the total in Vault increasing each time consumers swipe the card or conduct financial transactions. This feature is perfect if the consumer is trying to save up for something but always forgets to add to savings (Figure 9.4).

From Official Revolut Website

Digital banking **Revolut** is the latest FinTech to help you turn your spare change into savings gold (or cash or Bitcoin or any other major crypto).

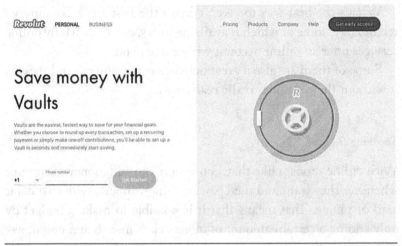

Figure 9.4   Revolut smart walet.

Revolut has launched Vaults, a new tool that allows users to set aside the spare change they receive from daily transactions. *"Every time you make a card transaction with Revolut, we round up your purchase to the nearest whole number and put the remaining money in your Vault"*, Revolut Chief Blogging Officer Rob Braileanu wrote this morning: *"Picture the scene - you buy your morning coffee for £2.70, we round it up to £3.00 and automatically put £0.30 into your Savings Vault"*.

Change set aside in Vaults can be taken out at any time, and users can adjust the amount they save so that more or less change ends up in their Vaults. Users of Vaults can save additional change in any of the 25 supported currencies and cryptocurrencies such as bitcoin, Litecoin, and Ether. In addition to converting change into savings, the app can also set up recurring payments to be set aside in your Vault or make one-time payments.

Loans of Tomorrow

One of the paths for Loans of tomorrow is crypto loans.

Getting loans with cryptocurrency is less complicated than getting bank loans – what exactly are crypto loans?

Figure 9.5 Crypto loans.

When investing, one of the most significant difficulties can be cash flow – and there's nothing worse than having to take the capital you've got tied up in assets for short-term costs and lack liquidity (Figure 9.5).

Let's imagine, for example, that Mark has 20 BTC. He doesn't want to sell it because he's confident that prices will appreciate substantially. Mark is also concerned that if he does end up selling his crypto, there will be a risk that ends up with less bitcoin when he repurchases it later.

Crypto lending platforms can come to the help here. Typically, Mark will be allowed to use his bitcoin as collateral – and get a loan in stablecoins. Owing to the volatility of digital assets, he will typically have to lock up more BTC than the overall value of his receiving funds.

Once he's repaid the loan, plus interest, his collateral BTC will be returned in total – and he'll make a profit if BTC ended up appreciating as he predicted. His BTC would only be at risk if he failed to keep up with the loan's terms or if the value of the BTC held as a collateral drop below the value of the loan he received.

When did crypto lending start? It was around the time that economies came to a screeching halt in 2020 due to the COVID-19 pandemic. This saw the interest rates get decreased and lending for big-ticket items take a nosedive. Crypto loans became an easy way to gain access to fiat currencies instantly, all without selling them.

Collateralized loans are more secure for the lender, enabling the borrower to take advantage of small interest rates.

Cryptocurrencies can be unstable, which is why these loans are overcollateralized. Overallocation provides insurance for the lender if the price of crypto fall. This can negatively impact the borrower – mainly if the landing platform they use requires them to keep their loan-to-value (LTV) ratio.

One of the biggest bonuses many see in a crypto loan is that, unlike traditional banking loans, clients will not be subject to credit scores being assessed. This means that lending is accessible to clients who don't have a known financial history, underbanked clients who don't have a bank account, and self-employed contractors who can't get credit because their fluctuating earnings don't meet a bank's lending criteria. Repayments are also more flexible.

The traditional old-fashioned financial world can take several days for loans to clear; BTC loans can be almost instant. The client can also make assets liquid without triggering a taxable event and adjust the loan to suit their needs. Clients can also switch between crypto assets to deposit Ether and borrow some other currency, all on the same platform.

Getting a Bitcoin Loan

There are two main options – centralized and decentralized lending platforms.

Centralized ecosystems such as Binance, BlockFi, and Nexo follow rules and procedures to be compliant. Clients have to create an account by signing up for chosen platform and go through Know Your Customer (KYC) and anti-money laundering procedures that are in place to prevent fraud.

These platforms have protocols in place to ensure that collaterals are safe. Best way is to protect crypto assets via insurance or keep the majority of the digital assets in cold storage, meaning they're off from an internet connection.

Centralized lending platforms will record all deposits and withdrawals using Blockchain technology, be transparent to everyone, and offer a great way to earn interest on bitcoin. The best savings accounts based on traditional fiat currencies offer interest rates around the 1% APY mark. Crypto platforms offer up to 8% on crypto interest.

There is a lot of paperwork involved in getting a loan through a traditional CeFi platform; there is a regulated environment – and a bank servicing agent could make these platforms more attractive to conventional investors.

The second option for cryptocurrency lending is to go via a decentralized Blockchain platform known as DeFi.

What Is a DeFi CryptoCurrency Lending?

DeFi lending platforms are entirely decentralized, and transactions are handled by code. On services such as Compound, Aave, and dYdX, smart contracts use protocols and algorithms to automate loan payouts.

Anyone interested in parties can access the logs on a decentralized lending platform, which makes them transparent, and nothing can be hidden on the Blockchain.

Unlike CeFi platforms, there is no intermediary or financial regulator, which means you don't have to go through a verification process like KYC and AML. However, DeFi interest rates for crypto loans often drop compared to what centralized competitors can offer.

Getting crypto loans on a DeFi platform is very quick as you won't need to pass due diligence. Thanks to smart contracts, all consumers will need to apply for the loan on a wanted platform and then send the crypto they want to use as collateral to a wallet connected with the chosen platform.

The clients of decentralized lending platforms can apply for a loan of any amount without confirming KYC identity. Loans can be paid in stablecoin, fiat currencies, or cryptocurrencies such as BTC or Ethereum.

Both types of lending platforms are still in their early stages. There's a lot of room for growth. The potential to access borrowing without the usual formalities could be a game changer, both for the consumers and the traditional financial services industry.

What Are Loan Models Available?

Investors can choose the loans which they want to finance. Platforms also allow lenders to declare how much bitcoin they are eager to lend and flexible interest rates. Borrowers can then ask for the loan that suits them best, and lenders can choose whether to provide the loan or not.

If a loan is denominated by bitcoin, the borrower must repay the bitcoin they borrow, plus interest. For example, if a client buy bitcoin and loans denominated by a primary currency (Euros or U.S. dollars, for example).

If a loan is denominated by bitcoin, the borrower must repay the bitcoin they borrow, plus interest. For example, if the client borrows one bitcoin for two years at 8% interest per annum, the client must repay one bitcoin plus 0.16 bitcoins as interest. This model works well if client earn or mine bitcoin because calculation can be made on how much the loan will cost.

If a fiat currency denominates a loan, the borrower must repay bitcoin to the amount of fiat currency. For example, if a client borrows 1,000 USD dollars' worth of bitcoin for two years at 8% nominal interest per annum, he must repay 1,000 U.S. dollars of bitcoin plus 160 dollars bitcoin in interest.

A more well-known use of this bitcoin loan type is an investment through which investors can sell bitcoin by holding the borrowed bitcoin and returning the agreed-on amount when its price increases.

Uses of Bitcoin Loans

The most significant benefit of crypto loans is that you can lend out bitcoins and get interests. You can keep your crypto to profit from capital gains if their value increases but still earn on interests.

As a borrower, the most significant benefit of bitcoin loans is that they allow you to get your hands on cryptocurrency. Investors can invest directly in services that require payment in bitcoin.

Bitcoin loans denominated by fiat allow investors to speculate if the price increase over the loan term. Borrowers can sell the borrowed bitcoin, repay the agreed-on amount of fiat currency plus interest, and take the difference in profit.

Risks of Crypto Loans for Borrowers

When loans are denominated by bitcoin but exchanged into fiat currencies and vice versa, the volatility of the value of bitcoin to fiat currencies is a risk. If borrowers earn standard fiat currencies and need to buy bitcoin specifically to repay loan, an increase in the price of bitcoin can increase the cost of loans. Drop in bitcoin prices reduce the cost of loans. In every case, getting bitcoin, which must then be sold to get fiat currency, and buying bitcoin to repay loans involves costs (including commission charged by marketplaces).

Risks of Crypto Loans for Lenders

There are a number of risks associated with lending out bitcoin:

Lack of Regulation

The regulatory landscape around digital assets like crypto, the development of which is in a constant state of flux, is insufficient. This can complicate debt collection when borrowers are not repaying loans.

In a Nutshell

What Will Be Good to Take from This Chapter?

Rise of Digital Currencies Central bank digital currencies (CBDCs) have begun to actively move up the agenda. By making digital currencies "official", they could change the face of the way money works. Digital coin backed by the central bank can support and improve the financial stability of countries. In China, a digital yuan could be an important step towards eradicating the grey and shadow economy.

Currently, cross-border payments involve high transfer fees. Central banks could programme funds in aid programs so that they can only be used for the aid or other purposes for which they were transferred.

In addition, there are important pushes such as the EPI, which aims to create an innovative and integrated payment system on the European Union and projects such as X-Pay in Germany to balance digital payment offers.

Data and Software Are the Heart of the Digital Transformation Economy The other major trend that is emerging is the development of new digital technologies that use data as a service. Increasingly, transaction data is being linked to additional key-value pair information to support the development of new digital services and leverage open banking – this is the thesis behind some of the biggest payments deals announced recently by credit card providers, cloud services, and online platforms.

It goes beyond traditional or retail/wholesale payments and connects business cases from the real economy that are linked to financial services.

Banks will compete on their data, leading to essential acquisitions of data capabilities to stay relevant in the digital economy. Skills, mindset, and culture will be as much a driver of M & M&A as the ability to innovate legacy technologies and corporate debt at scale, rather than making incremental changes from within that continue to be outperformed by the market.

Every bank must act decisively to find its niche. In a low interest rate environment with thin margins, they must find ways to maintain payment service profitability while avoiding being displaced by new, agile service providers.

What Is a DeFi Cryptocurrency Lending?   DeFi lending platforms are entirely decentralized, and transactions are handled by code. On services such as Compound, Aave and dYdX, smart contracts use protocols and algorithms to automate loan payouts.

Anyone interested in parties can access the logs on a decentralized lending platform, which makes them transparent, and nothing can be hidden on the Blockchain. Unlike CeFi platforms, there is no intermediary or financial regulator, which means you don't have to go through a verification process like KYC and AML.

However, DeFi interest rates for crypto loans often drop compared to what centralized competitors can offer.

Getting crypto loans on a DeFi platform is very quick as you won't need to pass due diligence. Thanks to smart contracts, all consumers will need to apply for the loan on a wanted platform and then send the crypto they want to use as collateral to a wallet connected with the chosen platform.

The clients of decentralized lending platforms can apply for a loan of any amount without confirming KYC identity. Loans can be paid in stablecoin, fiat currencies, or cryptocurrencies such as BTC or Ethereum.

10

INSURANCES BASED ON THE DECENTRALIZED NETWORK

If there is anyone dependent on your income – parents, children, relatives – you need life insurance.

– Suze Orman

How Blockchain Can Help the Insurance Industry

When it comes to reducing internal business conflict, smart contracts can be beneficial for facilitating and automating DLT networks. To this end, data reconciliation accessibility is improved. Furthermore, less time is needed to uncover information. This improves efficiency and transparency, creates a more seamless process, and cuts cycle times – all the while lowering costs.

Aggregate speed and accuracy improvements are pivotal. They enhance customer experiences, such as by reducing the claims cycle thanks to improved efficiency. This, in turn, enhances customer satisfaction and creates improved retention. It also makes the interactions between insurers and their customers more seamless and rapid; enhanced efficiency reduced costs, thereby lowering premiums.

Using Blockchain to Get the Basic Aspects Right

Two high-value areas of insurance are claims and finance functions. The Blockchain could be applied to both of these, especially in regards to processes that require ongoing reconciliation with external entities. For example, if Company A makes a claim against Company B and capital is subsequently exchanged by electronic transactions, the Blockchain could enhance efficiency.

At present, multiple insurance firms are using smart contracts alongside the Blockchain. This is triggered when the terms and conditions are adhered to. It's notable that the use of the Blockchain in this setting could create a contract whereby payments are made without needing human interactions. Hence, an insurer can process transactions more rapidly, leading to greatly enhanced customer service. Insurers can use the Blockchain this way to simplify and improve even complex processes.

Some ways that the Blockchain can be used to enhance the basics of insurance include:

- Enhanced subrogation processes
- Transparent claims processes
- Data-driven insights based on shared loss histories for prospective customers, allowing for more sophisticated pricing
- Increased efficiency during the claims process for payments between insurers and third parties

Using Blockchain to Support Innovations

Extended ecosystems are coming, and this will allow products and services to be provided more seamlessly. To this end, effective information exchange will be needed – which is where Blockchain comes in. The Blockchain doesn't need people working off of the same core system. Therefore, since the Blockchain allows for a more connected ecosystem despite not having the same admin systems or architecture, collaboration is much more secure and accessible. It's also more trustworthy; the Blockchain allows trust to be delegated to the ledger. In turn, this allows insurers to work together more freely to create microservices (commonly in APIs) to scale up digital partnerships. In short, the Blockchain will, in the long term, make it possible to create new, customer-centric products and business models.

The Blockchain can also power new business models based on personalized, real-time risk assessments. This differs from the current model, which relies on historical data and average pricing models. For example, this method of pricing could allow P&C insurance firms to build more sophisticated usage-based insurance models;

this could be done in collaboration with and working with support from vehicle or smart home device manufacturers.

That's not all. Some rapidly appearing customer-centric products and business models might include:

- Disintermediation
- Peer-to-peer (PTP) insurance
- Auto-adjusting policies
- Self-insurance
- Add-on of smart devices to existing insurance plans

Blockchain's True Value

Industry perceptions of Blockchain are rapidly evolving, and one of the most notable changes comes from the insurance industry. Indeed, insurers are now starting to look beyond DLT as being little more than an isolated enterprise technology. By contrast, insurers are beginning to where DLT's true value lies. In fact, DLT can serve as a catalyst for business ecosystem transformation.

This is similar to the way that cloud computing transformed the way we use technology, and it can be expected over the next few years to become a pivotal technology in industry disruption.

The majority of modern organizations have implemented the key digital age technologies: social media, mobile, analytics, and cloud computing (SMAC). However, the next stage of digital technology is rapidly becoming more powerful and influential. These rapidly growing technological innovations are DLT, artificial intelligence (AI), extended reality, and quantum computing (DARQ). Although each of these four technologies is at a different stage of development, they are expected to be key drivers for what Accenture refers to as the post-digital age.

SMAC technologies made it possible for businesses to use unprecedented levels of collaboration and connectivity. This was more than seen with insurance companies and their customers. Meanwhile, DARQ also represents a significant opportunity and has the potential to transform existing ecosystems, markets, and value chains – especially in the case of DLT. It is quite possible that business networks will be enhanced by distributed ledgers at never-before-seen speeds and scales.

In order to utilize DLT most effectively, insurers should partner with other firms to create DLT propositions and platforms. This is a rare opportunity to be one of the influential parties involved in shaping the future of technological solutions. In turn, this will help to create and enhance value for all members or participants of the ecosystem or value chain.

Investment in DLT

DLT and Blockchain projects have become massively more popular in recent times, and insurance firms are increasingly looking to get involved with this potential. In fact, according to a recent research report, the global market for Blockchain in the field of insurance is likely to grow from its value of $64.5 million in 2018 to $1.39 billion within five years by 2023. That represents a compound annual growth rate of 84.9%.

Research conducted by a company called Accenture also suggests that carriers are eager to implement these new findings. The Accenture Technology Vision 2019 survey found that over 80% of insurance firms have had some involvement with DLT; either their organizations had adopted DLT or were otherwise in the process of, or planning to, implement the technologies. Meanwhile, an additional study by Accenture and the World Economic Forum determined that 65% of insurance executives considered DLT implementation to be a vital part of offering competitive services to customers.

How Insurance Firms Are Implementing Blockchain Technologies

DLT and the Blockchain aren't brand new to the insurance industry and have actually been experimented with by insurance firms over the past few years in order to provide new insurance solutions for customers. These technologies are being utilized by all manner of insurance professionals, including insurtechs, incumbent insurers, and reinsurers.

As an example, Tokio Marine, a Japanese P&C insurer, trialed using the Blockchain for marine cargo insurance certificates. The company reported that using the Blockchain resulted in an 85% time reduction for insurance certificates to be received. Meanwhile, Allianz implemented Blockchain and smart contracts as part of their contract management, catastrophe swaps, and bonds processes; by doing so, they were able to make an otherwise complex procedure more seamless and frictionless.

The Blockchain can also be used to release value from other, locked-away areas too, such as the Blockchain-based proof of concept by Accenture. This proof of concept leverages data from smart sensors, thereby making smart-vineyard insurance feasible.

Using Consortia to Develop Blockchain Capabilities for Insurance Firms

B3i was a consortium founded in 2016 with support from brands including AIG, AIA, Allianz, Aegon, and Swiss Re. It focused on the insurance industry and was developing relevant Blockchain solutions. It then became a fully fledged, Switzerland-based company in 2018 and would become a key hub for Blockchain technology. In a similar manner, the R3 consortium includes in excess of 200 companies spread across six continents. The R3 consortium strives to reduce friction in transactions between financial services providers.

The Institutes is a well-known insurance education and certification body, and Accenture (serving as the lead framework architect) is part of its RiskStream Collaborative™, formerly called The Institutes RiskBlock Alliance. This is a Blockchain consortium that was created with the purpose of being for risk management within the insurance industry. Accenture's role is to assist with developing a production-grade platform able to build and, subsequently, implement Blockchain use cases.

The industry is currently under significant disruption pressure. Hence, the RiskStream Collaborative can assist carriers in uncovering methods by which to streamline their existing processes. At the same time, it could also help them to lower their risk portfolio while enhancing ecosystem efficiency and cost-effectiveness. Moreover, the Blockchain could also help with discovering new value chains, revenue streams, and operating models.

Over 40 Blockchain use cases have been identified at present by The RiskStream Collaborative. These would monitor and tackle core digital capabilities and innovation. As an example, preparations are already in place for the consortium to implement and test solutions for streamlining the first notice of loss. This could also be applicable for proof of insurance and subrogation cases. Moreover, it will be effective for helping to develop new smart-contract products.

Beyond Proof of Concept

Numerous insurance projects are now approaching production, and it is anticipated that these initiatives will refine their products over the next two years. Meanwhile, at this time, and as the industry catches on to the potential behind these solutions, further DLT insurance projects are likely to begin approaching production.

Two initial use cases for development have been identified by the RiskStream Collaborative. These are proof of insurance and first notice of loss. In the United States alone, by the third year of implementation, between $99 million and $277 million in annual savings are expected.

Several factors appear to be increasing the adoption of DLT within the industry. Firstly, software platforms including R3 Corda and Hyperledger Fabric are maturing rapidly. These are now at a stage whereby they can support production-grade DLT solutions. Secondly, growing numbers of carriers are using DLT to enhance efficiencies, despite growth forecasts being slower than usual. Finally, and potentially the most important point of all, key names within the industry are embracing the potential of DLT, assuming collaborations and partnerships are made to implement the solution. It's quite possible that DLT could the cost curve through revenue growth and cost reduction. As such, many industry powers are urging their partners, stakeholders, and customers to support the implementation of DLT technologies.

Successful Implementation of DLT within the Insurance Ecosystem

In order for DLT to be effective, it requires shared economic incentives and strong governance. Moreover, insurance companies from different ecosystems need to support the aforementioned features. This will enable insurance firms to implement an industry-wide solution that shares the cost of development while providing numerous benefits once fully operational. To this end, two primary models have been created: market leaders and peer networks.

Market Leaders

The market leader model requires DLT deployment to be primarily overseen by a predominant network activity driver in the industry. The aforementioned driver organization needs to be capable of driving efficiencies between business partners and customers alike. An example of this is Accenture and Zurich Benelux's Blockchain-based solution. It helps Benelux's insurers' customers manage surety bonds. Included in the solution is an intuitive and simple dashboard that allows customers to rapidly check information, including the status of their bonds, as well as finding detailed history records, viewing bond forecasts, and making new requests. The vision behind this project enables additional stakeholders to interact with one another. Thereby, this creates a more connected ecosystem without compromising data security and accuracy confidence.

Peer Networks

Organizations in similar industries converge via peer networks to enhance efficiency and value gains. This is relevant at a market-wide level. Examples of these insurance consortiums using peer networks include The Institutes RiskStream Collaborative and B3i. Both demonstrate how the model can be used flexibly depending on the size of the firm, with the initial emphasis being placed on creating a step change in efficiency for the entire value chain or market instead of driving competitive advantage for just a single participant. Therefore, additional participants will provide additional rewards for all members.

Insurance organizations need to create fair and structured incentives for all stakeholders involved, irrelevant of whether they aspire to lead a new consortium or build a peer network. All applicable governance structures and technology stacks must be created in such a way that they will be relevant for both the long and short term while also being scalable for large and small businesses a common set of interests.

Now is the time for action, regardless of which route the insurance firm decides is most appropriate. The World Wide Web and cloud computing came before and transformed the way in which we conduct business; similarly, DLT has the potential to bring considerable changes to insurance business' operations. By understanding DLT and its different industry applications, insurance companies can monitor and respond to both the challenges and benefits that are seen on the horizon.

Benefits of Blockchain in the Insurance Field

There are many potential benefits of implementing Blockchain technology in the field of insurance; these could include efficiency gains, cost savings, transparency, faster customer claim payouts, and fraud mitigation. Furthermore, the Blockchain will allow insurance firms to share data in real time between parties, all managed by trusted and traceable means. Moreover, there is potential for the Blockchain to enable the development of profitable new insurance practices and products.

Insurance firms are run in a field that is highly competitive. Indeed, retail and corporate customers alike demand the best value-for-money services from their insurance company as well as expecting more seamless online experiences. The Blockchain could enable growth and change that will benefit these demands.

Ethereum's smart contracts and decentralized applications make it possible for firms to conduct insurance policies over Blockchain accounts. For example, the Blockchain will allow for more automation as well as provide legitimate and secure audit trails. It is particularly notable that smart contracts and their transactions are highly affordable, which means that insurance firms may be able to offer more competitively priced products. This increases competitiveness and could potentially enable insurance firms to enter into the developing world's underinsured markets.

Cyber insurance would be required for the Blockchain, and this could be taken as a template for coverage. As part of this, the cyber insurance should cover the following: extensions and endorsements for financial loss, such as hot wallets and exchanges; specie and crime controls, such as cold wallets and vaults; professional liability for developers; and surety bonds to cover technology and software projects. By working alongside specialist tech companies, insurance brands will be able to reliably assess risk factors. Meanwhile, the technology companies will be able to advise insurers on best practices for loss control and mitigation.

Blockchain Insurance Uses

1. **Catastrophe Swap and Catastrophe Bonds**
   Insurers and investors could accelerate and simplify transactional processing and settlement through the use of contracts that are based on the Blockchain. This is notable as Cat swaps and bonds come with numerous risk factors, such as natural disaster risks.
   Cat swaps
   In exchange for payments, insurers pay third parties to take on the risk associated with catastrophe events. A catastrophe event could include, for example, a Florida hurricane. The third party would then be liable for any financial payments in the event that the catastrophe event should occur.
   Cat bonds
   With Cat bonds, a securitized financial instrument is used to share the catastrophe exposure between multiple parties involved. Interest is paid out in the form of coupons, if the catastrophe event does not occur, as well as the standard payout for bond maturity. However, if the catastrophe event occurs, the investors will be liable for making payment (Figure 10.1).

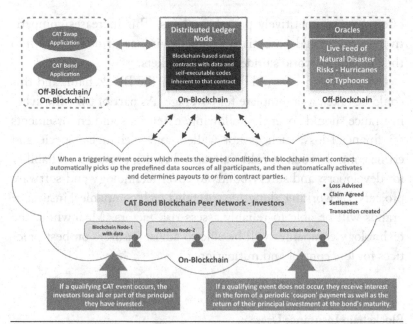

Figure 10.1 Blockchain peer network.

CAT bonds can be partially or fully automated using smart contracts.

Contract management processes could be readily accelerated through the use of Blockchain-based smart contracts, with every validated contract containing data and self-executable codes that are uniquely related to the contract. The Blockchain smart contract automatically detects any triggers meeting the defined criteria; thereby, automatic payments can be made. This enhances the reliability and auditability of the process; furthermore, it reduces the potential for frictional delays and human error. This also enhances the tradability of bond securities.

Network participants hold copies of the relevant ledgers. The system has the potential to enhance efficiency and reduce costs across multiple types of financial transactions. Furthermore, without the Blockchain, insurance firms use relational databases with unilateral consensus across different stakeholders. They can also deploy reconciliation, settlement, and audit services; this serves to create a consensus across stakeholders. As an example, following a claim being made, the transactions are processed and entered into multiple databases. Methods by which this is carried out include underwriting, claims, accounting, re-insurance and investors databases, etc. In the interest of ensuring that there is only a single version of the truth, these versions must be continually reconciled by the stakeholders.

The distributed ledger means that the Blockchain is able to break down large batch services. This is achieved by systemically incorporating services in a digital protocol-based consensus. The consensus is designed to include all of the transaction's counterparties and stakeholders; hence, all involved stakeholders have the same information and must use it in the same manner as the other stakeholders. Thus, the inefficiencies disappear in relation to relational databases, fragmented processing. This thereby provides a consistent, streamlined version of the truth (Figures 10.2 and 10.3).

2. **P&C claim settlement**
3. **Market investments**
   - Blockchain can be applied to market investments, which enhances transfer and fund administration efficiency. This is also applicable for financial collateral documents, including letters of credit, investment and collateral instruments, etc.
   - Transaction details for both the buyer and seller can be matched and agreement of the security price arranged. Moreover, transfer of ownership and associated rights and obligations can be documented and acted upon.
   - The Blockchain enables trade settlement, instantaneous report sharing, and conformance to rules and regulations. Moreover, transparency is also enhanced.
   - Cost and time reductions are implemented through Blockchain utilization.

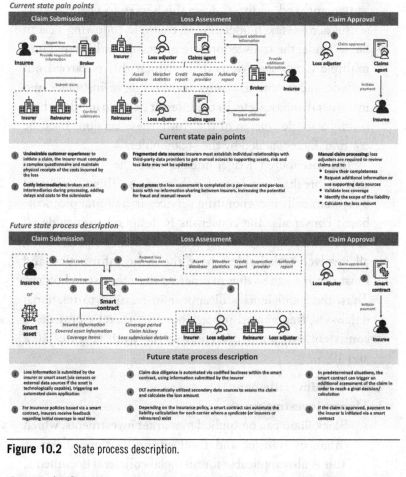

Figure 10.2 State process description.

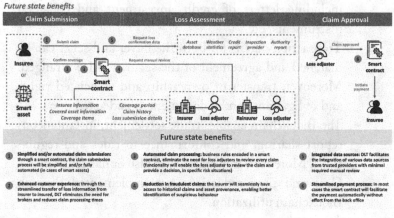

Figure 10.3 Future state benefits.

4. Financial audit and reporting

- Forgeries of the accounting systems are impossible or otherwise exceptionally expensive, owing to mutual control mechanisms, checks, and balances. Invariably, this has a significant impact on generic business operations. Some examples of systems that are currently manual but which could benefit from automation, include systematic duplication of efforts, extensive documentation, and periodical controls.

- The future of accounting systems could be based on Blockchain technology. Transactions can be recorded directly into a joint register instead of being individual records and input. Thereby, this creates an interlocking system of distributed and cryptographically sealed accounting records, for which falsification or destruction is effectively impossible.

- The process of standardization allows for automatic verification of important financial statement data, thereby saving significant cost and time for auditing. In turn, this would enable auditors' time to be utilized more profitably, such as for very complex transactions or internal control mechanisms.

5. Index-based Livestock Insurance Program

- Introduced in autumn 2015, the Kenya Livestock Insurance Program (KLIP) is intended to be rolled out across the 14 Kenyan countries.

- KLIP is an index-based livestock insurance scheme. It works through the use of the Normalized Difference Vegetation Index (NDVI), derived from satellite photography. The NDVI measures the photosynthetic activity of different plants, which is otherwise referred to as "greenness".

- The system measures the colour of the ground to obtain an indication of how dry the area is – on this front, yellow ground is very dry, while green ground indicates a good water supply.

- Farmers insured under the scheme automatically receive a pre-defined payment. This payment can be used to buy feed for their livestock once a predetermined dryness threshold is met. This payment is used to protect the livestock from the drought's effects but is not intended as compensation for the value of fallen stock due to the drought.

1. Flight insurance policy

- In some instances, due to the cumbersome and tiring paperwork involved with making a claim, passengers experiencing delayed flights did not make a claim, despite being covered under certain flight insurance policies.
- Smart contract solutions automate claim payouts. This is made instantly based on data provided by the flight and from verified flight data sources. This is completed via so-called "oracles", which serve to enable the use of external sources in the Blockchain.
- The claim notification step is rendered null and void since claims are made automatically through external verification. Hence, claims are closed more quickly, which reduces the processing costs while providing higher customer satisfaction.

2. Automate underwriting and claims handling

- Social network data can be used for verification of the employers' status for P2P supplemental unemployment insurance.
- Smart contracts serve to automate policy underwriting and claims handling. This is combined with approvals/verifications from other policyholders who serve as evaluators.

3. Internet of Things (IoT)

- Smart contracts can enable the creation of unique insurance policies for cars, electronic devices, or home appliances. The smart contract automatically detects any damage, following which it will trigger the repair, claims, and payments processes.

- Virtual currency payments are withdrawn from the policyholder's accounts based on the distance driven, which is calculated by the telematics system in autonomous vehicles, hence eliminating non-payment risks and providing automatic payments and premiums where applicable. Moreover, if the account is exhausted, the policy will automatically be terminated.
- The Blockchain can be applied for usage-based insurance during the registration of a policyholder's usage or health data. Smart contracts subsequently create or update the tariffs where necessary.
- A smart contract for crop insurance, when combined with Blockchain technology and crop sensors, could support the automatic settlement of claims for crop damage. When using the Blockchain and crop sensor systems, payments could be made before the farmer is even made aware that crop damage has occurred, ensuring optimal customer satisfaction.

4. Parametric insurance

- Fully automated insurance products are possible. In these cases, the contract is created using smart contract language. Meanwhile, the underwriting and claims processes are event driven based on data provided by real-world sensors.
- Certain insurance contract aspects can be automated through smart contracts, as driven by cognitive services and trusted Internet of Things (IoT) data feeds.
- Actual risk event occurrences (parametric events) result in payments being made. This is also triggered by certain index thresholds (e.g., hurricanes, rainfall, temperature, amount of available feedstock vegetation).
- Specific parameters result in a payment being made, so long as the parameter's value is communicated from the agreed expert (aka "oracle").
- Details of claims payments are automatically associated with the specific parameters that trigger payment for future contract refinements, updates to terms and conditions, and trigger parameters.

5. **Insurance of high-value assets (Ex: Diamonds)**
   - In the case of high-value assets, it can be beneficial to use Blockchain technology for cataloguing and verifying ownership. Companies such as Everledger, which supports diamond certification and related transaction histories, have emerged providing services such as these. These firms use Blockchain technology to provide cataloguing services to owners, claimants, law enforcement officials, and the like. It is able to monitor ownership of and track the location and movement of diamonds resultingly. Moreover, it can work alongside insurers or regulatory borders when diamonds are stolen, cross borders or enter the black market.
   - Individual and cumulative values are tracked through smart contracts and events, allowing for better coverage. Moreover, better risk management processes can be recommended; thereby, the customer can enjoy superior value overall.
   - Faster cycle times.
   - No-touch claims handling.
   - The enhanced information availability to establish suitable replacements may reduce the cost of claims overall.

6. **Reinsurance**
   - Blockchain has numerous potential uses in reinsurance applications, such as for recording specific claims' details to allow costs to be divided between insurers and reinsurers and providing an immutable record. Information detailed can include time-stamping when a specific claim was made, allowing a timeline to be created.
   - Shared Blockchains between insurers and reinsurers can provide a common view to contract details and associated financials. This enables a reduction in the incidence of disputes as well as easing reinsurance audits and smoothing cash flows. Smart contracts can automatically perform accounting adjustments as and when payments are exchanged.

7. **Processing of medical claims**
   - Multiple calls, partial payments, disputed coverages, errors in billing, and crossed payment/billing cycles are all common aspects of current claims processes. Moreover, the collection process regularly sees delinquent bills.
   - The shared Blockchain ledger captures and tracks changes made to the claim, and subsequently, smart contracts serve to confirm the applicable terms and conditions at various processing steps in the claims processing chain. By this method, it's possible for insurance firms to reduce the incidence of errors. The need for repeated work is hence reduced, while claims accuracy and traceability are improved.

Using The Blockchain is Most Relevant When the Claim:

- Involves numerous parties
- Involves new intermediaries
- Does not need a central trusted authority to execute transactions
- Requires an accurate record of each transaction's date and time
- Discourages retroactive manipulation of data
- Could involve multiple stakeholders using the same data

What Is an Electronic Signature?

Electronic signatures are electronic indications as to a person's intent to agree to the content of the relevant, specific document or a set of data. In much the same way as a handwritten signature offline, electronic signatures are legally binding and indicate that the signor is willing to be bound by the document's terms and conditions (Figure 10.4).

The Three Types of Electronic Signatures

As per the Electronic Identification and Trust Services (eIDAS) Regulations, three levels of electronic signature are defined. These are simple electronic signatures, advanced electronic signatures

Signatory   Document to be e-signed   Open a signing application   Authenticate (e.g. using a PIN) to access a qualified certificate on a qualified signature creation device (e.g. smartcard, remote service)   Electronically signed document

Figure 10.4 Qualified electronic signature.

(AdES), and qualified electronic signatures (QESs). Fittingly, the unique requirements of each level of electronic signature are based on a more complex variant of the previous level's requirements. Hence, by this logic, QESs generally have the most requirements, a simple electronic signature has the fewest requirements attached.

Simple Electronic Signatures

The definition of an electronic signature is "data in electronic form which is attached to or logically associated with other data in electronic form and which is used by the signatory to sign. By this definition, an act as simple as writing your name at the bottom of an e-mail could very well constitute an electronic signature.

Advanced Electronic Signatures (AdES)

AdES are based on simple electronic signatures but have the additional features that are:

- capable of identifying the signatory;
- created in such a way that ensures the signatory is able to retain control;
- linked to the document whereby subsequent data alterations are detectable.

Public key infrastructure (PKI) is the most commonly utilized form of infrastructure and is based on certificates and cryptographic keys.

Qualified Electronic Signatures (QESs)

A QES is an advanced form of electronic signature. In addition to the aforementioned aspects of AdES, it is:

- made using a qualified signature creation device;
- based on a qualified certificate for electronic signatures.

There are many different forms and varieties of signature creation devices, and these serve the pivotal role of protecting the signatory's electronic signature creation data. These could include smartcards, SIM cards, and USB sticks. Additionally, "remote signature creation devices" are used where the device is not in the physical possession of the signatory. Instead, it is managed by a separate provider. Users can benefit from enhanced overall user improvement when using remotely qualified signature solutions, which also maintain the same legal certainty as QESs.

Qualified certificates for electronic signatures are provided by (public and private) providers. Nationally competent authorities grant this qualified status, details for which can be found from the EU Member State's national "trusted lists" and accessed through the Trusted Browser List. Many qualified certificate providers use qualified signature creation devices in order to deliver the corresponding private key.

Different contexts require different levels of electronic signatures; however, it should be noted that only QESs have the same legal effect as handwritten signatures for all the EU countries (Figure 10.5).

When Should an Electronic Signature Be Used?

A variety of situations can require the use of electronic signatures. It should be noted that QESs are the most versatile since their legal effects are equivalent to a handwritten situation. Hence, they can

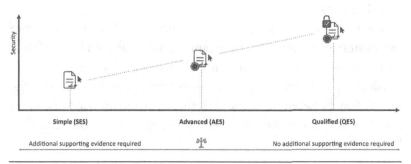

Figure 10.5 SES, AES, QES.

even be used across borders in situations where handwritten signatures could be used. These might include:

- Contracts
- Transactions
- Administrative procedures

Electronic Signatures in the EU

The Directive on a Community framework for electronic signature (eSignature Directive), which was officially adopted in 1999, was the first instance of legally recognized electronic signatures. Following this, electronic signatures in the EU became governed by the eIDAS Regulation as of 1st July 2016, which provides a predictable regulatory environment that is directly applicable to all the EU Member States. This nature ensures that secure and seamless electronic interactions are easily conducted between businesses, citizens, and public authorities in the European Union.

In a Nutshell

What Will Be Good to Take from This Chapter?

How Blockchain Can Help the Insurance Industry   Smart contracts can be beneficial for facilitating and automating DLT networks. This improves accessibility to data reconciliation and takes less time to uncover information. This improves efficiency and transparency, creates a more seamless process, and shortens cycle times – while reducing costs.

The improvements in speed and accuracy are key. They improve the customer experience, for example, by shortening the claims cycle thanks to improved efficiency. This in turn increases customer satisfaction and creates better customer retention. In addition, the interaction between insurers and their customers becomes more seamless and faster; the improved efficiency reduces costs and thus lowers premiums.

Benefits of Blockchain in the Insurance Field   Blockchain could be applied to both, particularly in relation to processes that require ongoing reconciliation with external bodies. For example, if Company A makes a claim against Company B and capital is subsequently exchanged through electronic transactions, Blockchain could increase efficiency.

Currently, several insurance companies are using smart contracts in addition to Blockchain. These are triggered when terms are met. It is worth noting that the use of Blockchain in this context could create a contract where payments are made without human interaction.

Some ways Blockchain can be used to improve insurance fundamentals include:

- Improved regression processes
- Transparent claims processes
- Data-driven insights based on shared claims histories for potential customers, allowing for more differentiated pricing
- Increased efficiency during the claims process for payments between insurers and third parties

Electronic Signatures in the EU   This nature ensures that secure and seamless electronic interactions are easily conducted between businesses, citizens, and public authorities in the European Union.

11
FinTech in Time

Banks have to upgrade themselves, or risk being burnt to the ground.

– J.P. Nicols

Approximately 12,000 FinTech companies exist globally, dominating the financial market with digital business models. Digitalization comes with numerous potential benefits and risks alike for banks and financial associations. Indeed, it could help to promote the bank's business model, but there is a concern that digitalization could reduce banks' market share and relevance. Hence, a balance needs to be found between ensuring the brand's products remain stable on the market while guaranteeing a secure gatekeeper position for the customer. Digital banking needs to be incorporated while still focusing on creating a multiple-channel/Omnichannel banking system that ensures good customer service and experiences.

Digital transformations can be a bit of a gamble, and this is something that IT financial institutions embrace. They have an open ecosystem, and this makes them excellent partners for everyday banking. Moreover, this provides the IT financial institutions with access to high-value, experience-focused customers who are willing to pay more for high-quality banking experiences. This could also include third-party providers who might be more interested in taking advantage of distribution channels to promote their products.

The banking market has long been covered by dominant technology firms, including Amazon and Facebook. These brands have seamlessly integrated these services with their customers' everyday lives. The leaders behind this digital banking solutions are Open Banking and Crealogix.

"The digital business model could represent the future of banking"

DOI: 10.1201/9781003198178-11

The Future of Digital Banking

Digital Open Banking Is Rewriting the Banking Rules!

There are a massive number of emerging technologies and business models that are changing the ways we go about banking and business. In a traditional sense, financial institutions created value through growth. However, modern value can be created through manifesting new opportunities through these emerging technologies. Moreover, external partnerships are also playing an increasingly important role in how the banking industry grows and changes.

In the traditional sense, banks operated on a closed ecosystem. Data sharing was limited solely to the customer, meaning that the services each brand can offer will be unique to the institution itself. However, the rise of shared ecosystems means that brands are able to provide personalized services for their clients. Naturally, this can represent a massive amount of potential for banking institutions.

Open banking represents significant opportunities for promoting the value chain of banking. Most notably, it alters how banks engage with their customers and offer services; indeed, open banking could allow banks to provide far superior services with more personalized services, working alongside customers on their own unique terms.

Open banking systems work by integrating third parties into the customer management processes. Application programming interfaces, otherwise called APIs, allow institutions to share software and data, which in turn can enhance the management and provision of services to customers. Notably, this allows customers to choose the right type of services for their needs.

In 2015, the European Parliament made it possible for banks to implement open banking systems through the revised Payment Service Directives, PSD2. This allows open banking solutions to provide real-time insights into finance management for customers while extensively using APIs for securely sharing data between banks. This also makes it possible for third-party developers to develop innovative services and apps within the bank's tech environment. Hence, this open banking situation leads to a collaborative ecosystem of financial providers.

We can define open banking as an ecosystem that furnishes the end user with data derived from a partnership of financial institutions derived through APIs, which make it possible for applications to collaborate and share information.

Indeed, APIs are the primary architecture of open banking ecosystems. By involving third parties in the open banking ecosystem, a scalable and secure API solution is developed that can provide a strong monetization opportunity. Moreover, APIs allow banks to implement ERP systems on behalf of corporate clients, while also performing e-KYC and PAN verification. This also ensures that the financial provider can offer currency rates, access and utilize credit scores, offer loans and other such products, work alongside third-party brands for the development of new products.

The extensive opportunity offered by APIs in the open banking ecosystem provide ample chances for FinTech companies to provide value-added services.

Some of the benefits of open banking ecosystems include:

- Superior transparency
- Personalized lifestyle integration
- Data visibility
- Customer profiling and product personalization
- Enhanced market efficacy
- Externalized customer-facing processes
- Potential to acquire and utilize new technology solutions such as the Blockchain or cloud
- Providing immediate customer interactions
- Third-party banking product features
- Leading innovation alongside partners

It's also important to consider the risk factors associated with open banking ecosystems. Indeed, open banking relies heavily upon the sharing of sensitive financial information between parties, and as such, data privacy is particularly of importance for consumers. Regulations such as the General Data Protection Regulation (GDPR) further increase the importance of ensuring careful data protection. Violations can be potentially damaging for the financial institutions, particularly from financial and reputational perspectives. All the while, it's important for banks and financial institutions to focus on value-added services and provide high priority to offering great customer experiences.

Four major roles of banking institutions for the open banking ecosystem include:

- **INTEGRATOR**: managing financial product production and distribution
- **PRODUCER**: creating new products and services
- **DISTRIBUTOR**: distributing in-house and third-party products and services
- **PLATFORM**: serving an intermediary role by managing both production and distribution services

Open banking in India represents a major possibility for banking firms. The Indian Government launched the Unified Payments Interface (UPI) which has paved the way for open banking in India. As such, this has been revolutionary in terms of retail banking.

Axis Bank began the process of investing in API services was AXIS Bank who would digitalize B2B supply chains. Following this, other banks including ICICI, RBL Bank, Kotak Bank, DCB Bank, and others would also implement similar strategies. Meanwhile, Aadhaar has seen over 878 million bank accounts being linked.

The number of transactions using Unified Payment methods rose to 312 million for August 2018, a 0.1% increase compared to October 2016. The value of transactions in this time rose massively from INR 0.5 million up to a staggering INR 542 billion, according to statistics derived from the India Stack website, National Payments Corporation of India, and UIDAI.

The Road Ahead for Indian Open Banking Ecosystems

It's safe to say that open banking ecosystems are being widely promoted in India, both by officials and industries alike, which is allowing for major non-linear growth. Thus, the customer is able to wield much more power over financial institutions, and solutions can be provided based on the customer's unique requirements and preferences.

From here on in, the focus now needs to be based on enhancing and supporting the ever-changing BFSI ecosystem. Steps have been made in this manner thanks to Aadhaar, eKYC, eSign, eMandate, and the like; however, there is still plenty of room for improvement in this manner, especially in terms of adoption of these technologies. Furthermore, educational programmes need heavy investment, in order to change public perception regarding the open banking ecosystem and digital economies as a whole.

Of course, for these steps to be effective, cooperation by financial institutions is required. As such, banks need to be willing to learn and change, and focus needs to be placed on ensuring that financial institutions are willing to embrace the changes ahead. The creation of new banking platforms and infrastructure will be pivotal in this innovation, particularly infrastructure that can safely and securely manage APIs without impacting the execution speed of requests for customers.

Financial Services Predictions

The financial services industry is forever changing, and it can be hard to predict how it will look in another decade. However, some predictions can be made about how the industry is likely to evolve. Indeed, as dictated by the "First Principles of Thinking", traditional models that have previously been a core architectural component of banking and rapidly being replaced by digital alternatives that enhance the user experience.

1. **2 Billion New Digital Customers**
   As time passes by, we can expect that approximately 2 billion (or more) new customers will sign up with financial institutions compared to 2010. Indeed, of these, approximately 1.4 billion new users will have never entered into a physical bank branch, with mobile applications being one of the preferred banking methods.
2. **Physical Branch Use Will Drastically Fall**
   The rise of the digital banking ecosystem means that more people will be using digital banking solutions, instead

of visiting traditional physical bank branches. Therefore, human advisors and branches will largely be reduced, with more customers choosing digital banking solutions.

3. **Sharp Increase in AI Advisors**

   Artificial intelligence systems will likely increase largely in the coming years, with AI helpers likely representing 50% of the total portfolio. As such, asset management AI and robots will likely represent a significant level of support for financial customers.

4. **Digital Banks Will Become the Largest Financial Institutions**

   It can be expected that, by 2030, digital banks such as Ant Financial will have grown in size to become among the largest financial institutions. Indeed, these institutions generally have lower costs and so may be able to offer more incentive packages for their customers.

Looking at the future of banking, it's safe to say that we will need to learn to embrace change. Indeed, although some changes may fall outside of our comfort zones, it will prove important that we are able to adjust to these changes. In fact, many features of our modern life would have retrospectively been unnerving when they were implemented originally.

Therefore, it's imperative that the ideal resources are allocated to the creation of the open banking model, including capital, materials, skilled personnel capable of implementing such a change, network resources, and the like. This will naturally represent a significant challenge for the financial institution, and the hassle may be off-putting for some banking firms.

However, it's imperative that these steps are taken if the banking industry is to grow to its fullest potential. After all, so many features of our modern life would have been missed if innovators hadn't taken the risk and taken the plunge! The immense changes that have occurred in the last decade alone are a testament to this fact, with many of the technologies that we use on a daily basis today potentially were not existing just over a decade ago. Key examples of this include Uber and Lyft, two immense brands that have massive influence in how we live our lives today. Yet, these two brands were founded in 2009 and 2012, respectively, and so are still relatively new in comparison to many other institutions. Meanwhile, websites such as Snapchat, Instagram, and Pinterest are even younger, being developed in the mid to late years of the 2010s.

Of course, technology and the embracing of new technological innovations have been pivotal in terms of these developments. However, the banking industry still has a large amount of potential for advancement. Some recent innovations including robo-investments, lending Tech, cryptocurrencies, and mobile/online banking have been hugely successful. Therefore, it's absolutely vital that banks take the opportunity and serve as active contributors to the open banking ecosystem.

Predictions

Banks Will Become Similar to the Post Office in Terms of Perception

Most of us do not directly deal with the post office, and instead, we tend to only handle the post office through other brands and multiple tech layers (for example, after placing an order with Amazon). The same will likely be the case with banking solutions, too. FinTech development will create a new layer between the bank and their customer, and as such, this will have a significant impact on how consumers deal with their banks. Indeed, Gartner predict that 80% of traditional banks will have a different relationship with their customers in the future, and the remaining 20% will be one of the following four types of firms:

- FinTech infrastructure providers which are capable of rapid and affordable scaling and product development
- FinTech technology developers for discrete products
- Traditional financial institutions implementing digital platforms, particularly targeting low-value customers in the working class or lower-class niche
- Concierge providers specializing in bundled offers for high-value customers

All Businesses Will Be FinTech-based

A second prediction is that all businesses will have to evolve into a form of FinTech businesses, either by launching FinTech products or otherwise delivering FinTech inspired financial experiences, such as with Apple launching its Apple Card and Walmart launching the Cashi payment system. This is due to the immense potential of "infrastructure as a service" models, which have allowed for in-house financial products to be much more efficient, effective, and beneficial for consumers.

As explained by Richie Serna, the co-founder and CEO of Finix, *"financial services will be a core ingredient of every software company. In the same way we don't talk about "internet" companies anymore, we'll stop talking about "FinTech" companies".* Finix, in particular, is a valuable example of how businesses could create their own in-house payment methods. The brand provides technological support for brands in all niches to accept payments through their own payment processor systems, instead of needing to rely on third-party payment solutions such as Stripe or Square.

Latin America Will Show the Fastest Rate of Growth

Somewhere in the region of 70% of Latin Americans are underbanked or unbanked. The number of Latin Americans owning a smart phone device is expected to reach 78% by 2025, compared to 64% in 2018; this represents somewhere in the region of $34 billion in value.

As a result of this, FinTech providers are implementing new methods of payment for an otherwise informal economy. This has led to a staggering investment of totally $2.1 billion across 130 FinTech deals for 2019 alone. Moreover, other countries are also seeing similar levels of investment. This is further enhanced by the implementation of laws in Latin American countries that ensure FinTech brands and traditional financial institutions are equally regulated.

The War between Banks and Aggregators Will Be Lost

Recent times have seen a growing trend of bank aggregator models, such as apps allowing customers to directly link their bank accounts in order to allow for more seamless financial management.

Commonly, this is carried out with the use of third parties such as Plaid, Yodlee, and Finicity, which rely on web scraping tools to interact with the banks – and naturally, this has caused a notable degree of friction due to security concerns.

However, a study by Accenture found that over 70% of customers would trade some degree of financial security for simpler financial management, and as such, it is likely that banks might lose the war over the rights to consumers' data. Legislation dictating that consumers own the rights to their data further support this prediction, and so, it seems to be only a matter of time until this is drawn to a close.

Ant Financial's Market Share Will Grow to Make it the Largest Financial Institution Globally

Ant Financial is a spin-off of Alibaba and is one of the most valuable banking providers globally with a total of 1.2 billion users utilizing its wallet application. The app offers a plethora of different features that make it potentially more beneficial for consumers than standard banking systems including mobile payments, savings and investment accounts, lending, and credit scoring. These services are intended to make financial management easier for customers.

Moreover, the brand also provides a platform from which other financial providers also work, with over 200 partners and financial institutions being located globally in Asia, Thailand, the Philippines, Pakistan, Bangladesh, Malaysia, Korea, and Britain.

As a result of this, it is a hugely valuable company, with its predicted value in 2018 being $150 billion. Moreover, in the third quarter of 2019, it was responsible for increasing Alibaba's bottom line by an impressive $309 million.

Hence, it can be predicted that this trend of growth will continue. If Ant Financial should achieve its growth plans and enter the rest of the developing world with its financial products, it could surpass rival financial firms and become the largest global financial institution.

Customers Will Avoid Physical Banking in the Future

As time passes by, it's becoming increasingly apparent that the trend towards digital banking solutions is reducing the reliance on traditional banking and physical branches. The COVID-19 pandemic has further enhanced this, leading to more and more people managing their finances virtually as opposed to in a physical branch.

As such, when the rate of growth for digital banking is considered, an important question needs to be asked. Will physical banks have any role in the future?

Annually, 2,000 national retail banks have closed over the last three years. As such, it's quite likely that this will continue, and the improvements to digital banking many further promote the value offered by digital solutions instead.

Advancements in Financial Technology

87% of consumers now use mobile banking applications, and this increases to 97% of millennials. As such, it can be expected that the use of branches will continue to fall alongside the advancements in financial technology. Indeed, with so many consumers now preferring digital solutions over traditional branch banking, it seems that the move towards digital banking is increasing rapidly. Digital banking solutions allow users to check the balance of their accounts, transfer funds, deposit money, and open new accounts. Moreover, the recent advancements that have been seen in terms of wearable technologies, such as smartwatches and mobile wallets, have also fuelled the advancements we have seen recently in terms of the popularity of digital banking solutions and the fall of traditional branches.

Now, only 20% of consumers are likely to use a branch for their banking needs. Instead, many consumers prefer to use digital banking solutions for their financial management and payments. The frictionless nature of digital banking is largely part of this.

Growth of Digital Banks

The rate of growth of digital banking solutions is massive, but digital-only banks are also becoming more significant. Indeed, increasing numbers of banks are opening without having any physical branches and only having digital solutions for consumers. Banks such as Monzo and Chime fall into this category, and they are especially consumers for the increased financial incentives that they can offer for their customers due to not having physical banks. Moreover, these forms of banking models are commonly more integrated with our day-to-day lives, with many featuring immediate fraud or low-balance alerts. The only limitation is that some customers may choose standard banking options for the security that a local branch can offer. Even those who do not necessarily want to use physical banks tend to prefer having a local branch as a form of security.

Concept Branches

Generally speaking, most physical branches for banks have been falling out of favour. However, there is one form of branch that is becoming more popular, and that is the concept branch. Concept branches, such as Capital One, are a somewhat experimental form but they are proving to be popular and effective. They are run across multiple regions of the country, such as in coffee shops, and provide a suitable option for customers who want a physical branch without necessarily needing to rely on these on a regular basis.

Branches for the Unbanked

There are a huge number of people globally who do not use banking solutions still, and it is hence likely that physical branches will never be entirely wiped out. Individuals who rely on cash for their financial management, such as those individuals living in poverty or with low incomes, will be unable to benefit from online banking methods. As such, physical banks are still important for these individuals, with as many as 78% of Americans living pay cheque to pay cheque. Furthermore, with cities banning cashless society for businesses for this very reason, it is important that physical banks still remain available.

As such, the physical location of branches is now largely becoming dependent on where they are needed most. This trend is likely to be one that continues in the future, with physical branches being carefully positioned in locations where they will be utilized most effectively.

How Banking Will Change by 2030

As we progress forward with digital banking solutions, it is almost inevitable that the way we bank will be transformed significantly. Indeed, although it is currently unlikely for society to become cashless in the immediate future, we are spending much less hard cash than before. As such, digital banking is becoming much more prominent.

This is being seen in numerous ways. The maximum spend for contactless payments, for example, has been increased. Meanwhile high street branches of traditional banks are closing in favour of digital banking solutions. As a result of this, banks are able to save money and pass these savings onto their customers. All the while, more people are opening up investment accounts, with many banks adding hundreds of thousands of new clients for this purpose.

As such, while it's impossible to predict how the future of banking will look specifically, we can come up with a few accurate ideas about what to expect.

People Having a Say

The banking industry is facing growing pressure to become more user-friendly, and this has largely driven the drive towards digital banking and mobile banking. As part of this, consumers are increasingly demanding more customized and personalized banking solutions – and that's where banks such as Monzo and Starling are really excelling and leading the way, providing interactions based heavily on promoting ownership of the funds. Interactive banking and tailored solutions for individuals further help to promote this, with many schemes being aimed at enhancing the value that customers get from their banking provider.

Diversification is an essential part of ensuring that the banks of the future are effective, but traditional banks need to ensure that they have the technologies in place soon for this. Indeed, rival brands such as Monzo are rapidly taking control in these customer-centric models, and this could somewhat threaten the legacy banking model if swift action isn't taken.

"Best of Both" Business Models

It is likely that, by 2030, banks will be working together in a more collaborative manner than they do today. Indeed, while digital platforms can provide efficient customer support and cheaper costs, traditional banking systems largely are more powerful and can provide superior investment and lending capabilities. By working collaboratively, these brands may be able to provide services that are more suitable to customers' needs without having to choose one extreme or the other; as such, physical and digital banking solutions alike should be available for customers to choose between, creating a seamless experience without compromising on security or accessibility.

Increased Security and Growth Focuses

It can be difficult to find the delicate balance between security and efficiency, however, it is likely to be the case that banks and financial services providers will have found some degree of compromise between these by 2030. Improvements in cybersecurity should ensure that digital banking solutions don't compromise customers security and data privacy. Meanwhile, further investment in technology and artificial intelligence will hopefully drive significant improvements in terms of threat detection and management. False positives should, as such, be largely ruled out, and this will play a significant role in the reductions of fraud and money laundering cases.

Regulation will likely have become much more pivotal by 2030, too. Indeed, this is something that we are already seeing, with central regulators taking a firm stance on the potential risks of their business models and taking steps to mitigate risks where possible. Furthermore, fast-tracked programmes should be effective for helping to strategically reduce risk while enhancing the potential for monetization and streamlining of financial activities.

Innovation

Invariably, by 2030, banks and financial providers will have spent a lot more time and money on innovating the products that they currently offer. As such, although this may increase the level of competition between providers for a while, it is likely that we will see some increased collaboration between providers to offer the most innovative banking solutions to entice new customers. These innovations will largely be driven by the customer's needs, with banks implementing strategies that best suit their customers' requirements while also focusing on ensuring profitability for the long-term basis.

In order to achieve this, it is likely that investment will need to be made into new teams dedicated solely to product and innovation development. A chief innovation officer, for example, could become a highly influential role in the board room for the future of banking, backed up by a team of experts who carefully analyse every customer's expectations and critiques in order to tailor a product and banking solution that is better suited towards the financial market of 2030. Of course, innovation will continue to evolve long beyond 2030, so this will be a change that will pick up pace and stay, providing great value services to customers while keeping banks profitable and secure.

Future innovations will, as such, be customer-driven and targeted towards enhancing the customer experience of financial customers overall.

Enhanced Security Focus

In the past, banks were largely based on managing finances and little more. However, modern financial institutions are facing increasing pressure to carefully manage user data, too, and they have become custodians for our personal information in modern times. As such, it's equally becoming increasingly important that banks have effective security solutions in place – likely powered by artificial intelligence and improvements in technology.

As argued by KPMG, by 2030, it is likely that a considerable sum of financial institutions' profits will be comprised from running Machine Learning analytics based on users' banking data. In doing so, this will make it possible for the data to be secured, managed, and subsequently distributed on behalf of customers. KPMG explain, "Banks will manage your data like they do your financial assets… Consumers will understand the value of their data and demand a return". In other words, these changes will likely transform data into a form of currency, of sorts.

This would be beneficial for a selection of reasons. For one thing, it would make it possible for banking providers to readily share relevant data and personal information with service providers, saving time from the customer's perspective. What's more, this data would allow banks to provide their customers with highly customized services that are directly tailored based on the customer's unique preferences and requirements.

Emerging Regulations for Technology

Regardless of whether the previous predictions for the banks of 2030 prove to be correct, it's a safe bet to say that the banking industry in 2030 will be one that's vastly different from the industry we know today. Indeed, this will be driven by a number of different factors, including increasing pressure from customers to provide tailored services, a reduction in reliance on physical branches, and potentially emerging regulations for technology too. Indeed, regulations such as the privacy GDPR regulations in Europe have major implications, both good and bad, for the banking industry.

Furthermore, regulations for anti-money laundering are helping to support the provision of premium quality services for customers, while encouraging financial institutions to further invest in this highly important field of security.

What's more, new technologies are also having a big role on how the banking industry is changing. For example, 5G wireless technologies could open up the world of digital banking to an even more massive audience with new types of wearable technologies potentially being on the edge of existence as a result. And need we even get into biometric technology? These are just a selection of the different technologies that could revolutionize banking as we know it. All are closely monitored and regulated – and as such, they could provide a highly effective solution for ensuring that the future of banking is dynamic, effective, seamless, and available for everyone.

Big Tech and Borrowing

Massive corporate data breaches are occurring on a worryingly regular basis across some of the world's most influential brands and companies. As a result of this, it's no surprise that many people have concerns about the safety of their data, even with regulations such as GDPR theoretically protecting us against data breaches of these sorts. As such, it's important to consider whether we as the consumer are ready for banks and financial institutions to take over the management of our data fully.

As explained by Luke Thomas, futurist in the CTO's office at HP Inc. in Palo Alto, California, *"It really depends on what you consider to be a bank"*. In other words, it's important that we turn this question around and look at it another way.

Indeed, according to Thomas, other big brands such as Amazon, Google, and Apple will likely begin to provide their own banking and financial management portfolios; as such, he goes on to explain that,

> these Big Tech 'banks' will have so much data on us, gleaned from so many touchpoints, that they will indeed be able to help us control our data and manage our money in much more efficient ways. I don't mind sharing my information as long as I know how it is going to be used, especially if I get compensated for it. And with GDPR we'll get opt-in options, creating a marketplace based on trust.

This opinion is very similar to the predictions made by KPMG, so it's something that is important to consider. As such, while it might be that banks have a role in data management in the future, the banks themselves may not be quite the same institutions as we know today.

Social Media Currency

At present, Facebook has a plan for starting its own form of cryptocurrency, and this could potentially have an impact on the future of banking systems. Indeed, the threat posed by social media currency for banking solutions is one that cannot be completely ignored. As such, this could have an impact on the way the banking and financial management industry looks by the time we reach 2030. Indeed, with a platform of over 2.2 billion users around the world, the currency on the platform could represent a significant change to the way we spend online. Indeed, the potential of the Facebook-powered Libra cryptocurrency is such that the world's biggest central banking authorities have shown serious alarm – and so it's worthwhile to consider it as part of the future of banking.

"With 2.2 billion users, the moment Facebook gets its global regulatory approvals it will instantly become the biggest bank in the world. And the biggest financial institution in the world", says Thomas. What's more, he also notes that this move would give Facebook massive insight into how people spend their money, making them one of the biggest global controllers of data too. However, it is worth noting that many users could find this to be intimidating, and as such, there could be potential for banking institutions to work with this to their advantage.

So, will social media currencies represent a major shift in the banking and financial industries by 2030? This will likely depend on whether Libra is able to attain critical payment system partners. Without these partners, including Visa, Stripe, Mastercard, Ebay, and PayPal, the Libra currency will likely fall flat; as such, the success of social medial banking will be largely dependent on how well Facebook is able to partner with payment processors.

However, the potential risk for users to be scared off by privacy issues could have a significant impact on the success of the Libra cryptocurrency as well. With that being said, this may not necessarily be something that users spend too much time worrying about. As explained by Thomas,

Facebook users might decide that because their data is already exposed on the internet, and as they are getting a nice deal with Libra, they'll go for it. And bear in mind that whilst privacy issues are a big thing in the west, people do not care much about it in emerging markets and Asia, our research shows. So Facebook could still become a big financial institution in those markets.

The Impact of Social Banking

One way or another, the fate of the Libra cryptocurrency will likely be decided by 2030. Indeed, Jared Dame, Director of AI and Data Science at HP Inc., explained that

> Banking a decade from now will be much more about forging a human connection. It will be about developing a relationship between the bank and you, the customer: the bank's AI will really know you. They'll know about your spending habits, your savings and credit lines. And it will ask you about that car you bought, that horse or that boat. It'll even ask you how your partner and kids are doing.

Indeed, the future of the banking industry is set to be largely reliant upon customer experience, and this means that many banking firms need to focus carefully on the experience behind their services. As such, while this might sound somewhat unnerving for some customers, a hyper-personalized approach will likely be a pivotal part of creating the bank of the future.

Indeed, Jared Dame believes that, when the IT systems behind the bank are largely optimized and automated, having the human touch of personalized banking services could be pivotal in ensuring that services remain friendly and amenable. In contrast, though, Thomas' opinion is that society doesn't necessarily crave that human interaction from their bank providers anymore. As he explains,

> You won't want to wear heavy AR or VR headsets. You can do a lot of this stuff on your phone with a chatbot or virtual assistant. The fact that branches are closing down suggests people do not want that face-to-face interaction.

As such, over the coming decade, we will likely see which of these approaches is the more appropriate.

Can Blockchain Provide Security?

At this point, we need to consider whether security can be provided by the Blockchain and quantum solutions. Indeed, in a rapidly evolving digital world and with the arrival of 5G upon us imminently, it's safe to say that technologies such as Blockchain-based digital ledgers will have a significant impact on how the future of the banking industry develops.

As Thomas explains, the Blockchain is especially useful from a security perspective as it cannot be hacked and is not controlled by any central authorities, providing an extra layer of security and confidence for customers.

Furthermore, the Blockchain automatically encrypts and flattens data into linked blocks, thereby providing a record that would require hackers to possess a majority (51% or greater) control of the network to even have any chance at succeeding with a successful hacking attempt.

As such, it is ideal for storing unchangeable (immutable) data, especially when used in combination with superfast 5G-based internet networks. The quantum computer does somewhat challenge this security, though.

The quantum computer, as the name would suggest, is able to utilize quantum particles to speed up the process of completing calculations. Thus, it could complete calculations far faster than any standard computer, even the most powerful models. So, quantum computers – if successfully developed and implemented – could endanger the success and security of the current Blockchain.

How powerful is the quantum computer, potentially? Recent research by Google and Nasa determined that the quantum computer could be so much faster than normal computers, that a 200-second quantum calculation would take a staggering 10,000 years to be completed on a top-of-the-range modern computer device. However, these forms of quantum computers are still in their infancy and could equally be used in the alternate sense to provide impenetrable cyber defences.

"It is always good to differentiate between hype and reality. Yes, quantum computing could one day break some of the encryption we depend on – but we are many years away from it being mainstream", says Thomas. He then goes on to explain, *"Quantum computing could provide us with strong cyber defences. It just depends how much you want to spend"*.

The Fate of Laggard Business Models for Banking in 2030

As with all industries, it's always essential that business models evolve. To this end, it is inevitable that data will play a significant role in the future of banking in a decade's time. Head of KPMG, Jonathan Holt, says, "Financial services will become vastly more interconnected and collaborative – and virtually frictionless. Data will fundamentally change the business of financial services and it will be at the heart of how financial services will make money in 2030". However, his concerns revolve around certain long-established banks, and he finishes by saying, "Many financial services CEOs are not embracing the simple fact that their very business model will need to change. And fast".

In Conclusion

We are living in the era of the FinTech industry for decades, even centuries. Sir Isaac Newton, as a Master of the Royal Mint, started the industry when he applied scientific principles to recognize counterfeit currency. We can then say that the FinTech progresses back from 1696 and his time.

However, we define that the last decade has seen FinTech impulse's unprecedented development in financial services. Billions in investment, new businesses with services, and ways of offering those services. The FinTech sector is also emerging, with competition giving way to collaboration as banks and big tech form uneasy relationships.

As we entered a new decade, the future of the FinTech industry is at a crossroads. It's reasonable to recognize what the FinTech industry will look like in 2030. Will banks remain the most prominent players or will grown-up tech overtake them? Will data replace transactions as the primary revenue stream for the industry? Has innovation flourished or just getting started?

FinTechs are not predicting banks' death, of course. But there is a profound reflection on what the long-term future of the FinTech sector looks like, with many seeing a future where financial services will be deeply embedded into all technologies — and therefore our daily lives — that the term FinTech becomes essentially insignificant.

Today, it's almost unnecessary to use the term "online business" or "internet business". Every single business has some online presence if they want to do business at the beginning of this decade, even if it's just a Facebook page or Google ads, doing every business online.

Perhaps, Blockchain, while gain importance in cooperation with open banking and emerge necessary changes that are expected. Machine Learning and automation are predicted to have a significant impact on the banking sector. The growth of Machine Learning powered by Quantum Computing and AI will lead to data analytics becoming the most important currency of tomorrow.

Printed in the United States
by Baker & Taylor Publisher Services